BRITISH PALAEOGEOGRAPHY

An Introduction

BRITISH PALAEOGEOGRAPHY

An Introduction

by

P. R. RANKILOR

Altrincham

JOHN SHERRATT AND SON LTD

First published 1972 by
John Sherratt and Son Ltd.,
Altrincham

1SBN 0 85427 028 0

Printed in Great Britain
at the St. Ann's Press, Park Road, Altrincham

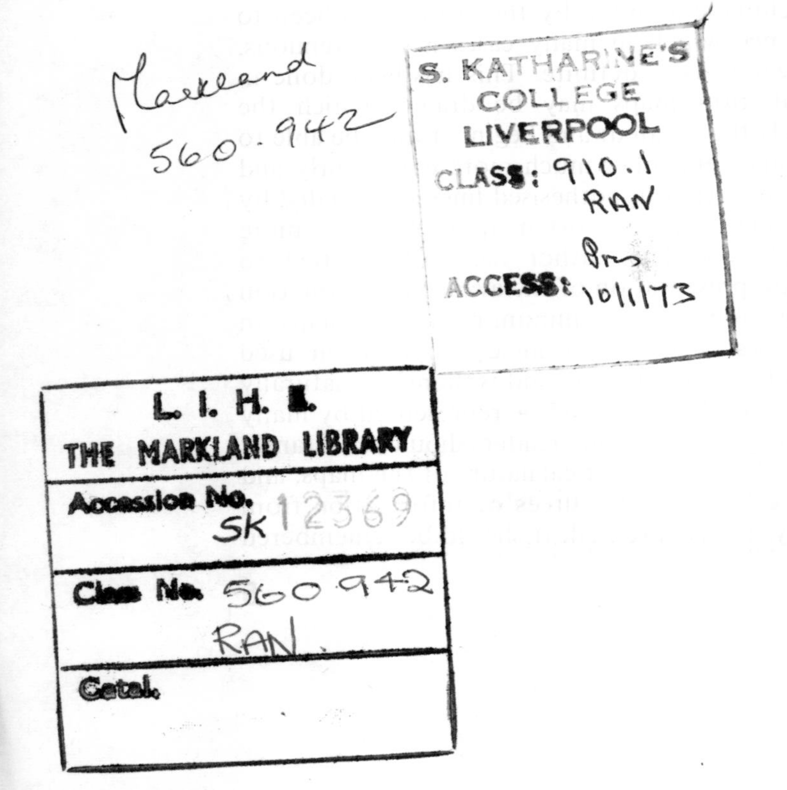

PREFACE

The object of this book is to give students, and others interested in geology, an introduction to the study of the Historical Geology of the British Isles. The text is presented in the form of a map-study note book and covers a time period of some 600 million years from Cambrian to the present day, and a short mention of the Pre-Cambrian is also made. The main fossil types found in the sedimentary rocks of each period are mentioned, along with the general names of the major rock groups.

The technique adopted by the author has been to take evidence which in many cases is only tenuous, and to treat it as if definite. This has been done in order that firm maps may be drawn, which the author feels the amateur and beginner may be able to recognise and remember much more confidently and swiftly than dotted hypothesised lines surrounded by questions marks as are often appropriate in more advanced books. The author has further tried to simplify the presentation of the available information by the selection of the minimum number of maps. In as many cases as possible, a single map has been used to represent an entire period, and is therefore naturally a compromise of what could be represented by many maps. It is vital that the reader should constantly bear in mind the hypothetical nature of the maps, and the often controversial sources of information from which they are constructed. It should be remembered that very broad generalisations must be made to produce a map which is supposed to represent the geography of a period which may extend up to one hundred million years in length. These maps, therefore, are intended to serve as a broad pictorial base upon which the beginner may build a more detailed understanding of one of the most interesting aspects of the Geology of the British Isles.

A great deal of the information used to construct these maps has been extracted from "A Palaeogeographical Atlas" by L. J. Wills, 1951, Blackie, London; and also "The Geological History of the British Isles" by G. M. Bennison and A. E. Wright, 1969, Edward Arnold (Publishers) Ltd., London. The author wishes to express his sincere thanks to the authors and publishers concerned for their permission to make use of such information as was required. Both of the above publications are authoritative in their fields, and by ensuring reasonable compatibility with the above, the author hopes that interested beginners may proceed to further detailed study with the maximum ease.

P.R.R.

INTRODUCTION

The maps presented in the following pages cover some 600 million years of the Earth's history. Their purpose is to try and indicate the major changes in geography that have taken place during this time. Geography, in this case, is a term which encompasses the relative positions of land and sea; the prevailing climatic conditions; the direction of prevalent winds and ocean currents; the type of life present in the sea and on the land; and even the fundamental position of Britain during its migration away from the Southern Hemisphere.

A map can provide an immense amount of information to the reader in a very short time; indeed it can describe what words alone would take many pages to explain. Working on this basis, the author has attempted to keep the written word to a minimum, and to convey as much information as possible through the maps.

All the maps cover an entire period or more, and are accompanied by simple descriptions including the fundamental sub-divisions of the rock groups within that period. Relevant fossil types are also mentioned.

In general wherever land is shown on the map, the reader may assume that rocks were either being eroded, or terrestrial deposits were laid down. In the case of the Devonian and Permian/Triassic Periods, desert sandstones were being formed in the lowland desert regions whilst erosion was taking place in the highland zones. Wherever the sea is indicated the reader may assume that marine sedimentation was taking place.

The well-known "10 miles to the inch" map of Britain published by the Institute of Geological Sciences, indicates very clearly to the student the present positions of the major rock groups in this country. It is hoped that some understanding of the geography at the time of deposition of the various groups will give the student an idea of which rock types to expect in any given area, and will also increase his enjoyment in observing these rocks by giving an understanding of their mode of origin, and relationship to the other outcrops of the same period, but in different parts of the country.

For a further reference on stratigraphy, the student is recommended to "A Stratigraphy of the British Isles" by Dorothy Rayner.

There have been, in recent years, tremendous advances in the understanding of continental drift, mountain building, and the construction of the Earth's crust. New concepts such as "plate-tectonics" and "ocean-floor spreading" have been put forward, and are becoming widely accepted. To those students who are interested in further reading on this subject the author recommends the Open University Set Book "Understanding The Earth" edited by I. G. Cass, P. J. Smith and R. C. L. Wilson.

TIME SCALE

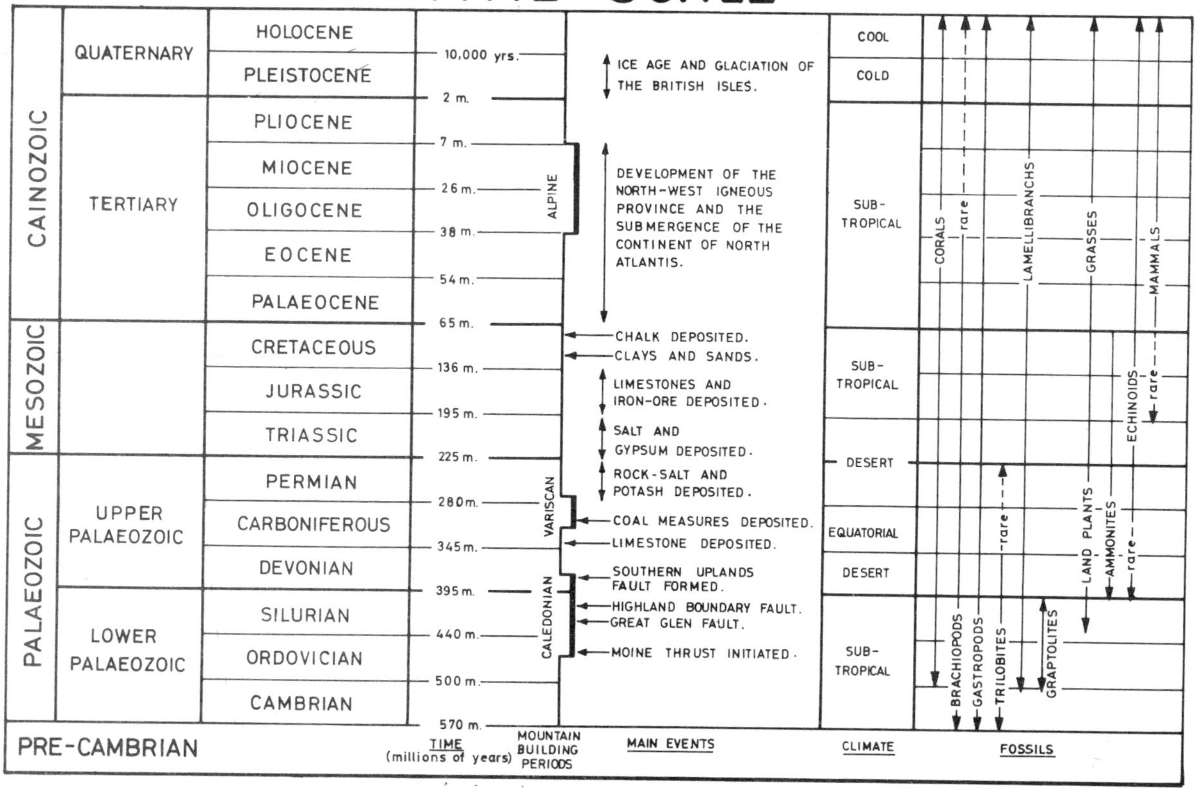

Era	Sub-era	Period / Epoch	TIME (millions of years)	MOUNTAIN BUILDING PERIODS	MAIN EVENTS	CLIMATE
CAINOZOIC	QUATERNARY	HOLOCENE	10,000 yrs.		ICE AGE AND GLACIATION OF THE BRITISH ISLES.	COOL
		PLEISTOCENE	2 m.			COLD
	TERTIARY	PLIOCENE	7 m.		DEVELOPMENT OF THE NORTH-WEST IGNEOUS PROVINCE AND THE SUBMERGENCE OF THE CONTINENT OF NORTH ATLANTIS.	SUB-TROPICAL
		MIOCENE	26 m.	ALPINE		
		OLIGOCENE	38 m.			
		EOCENE	54 m.			
		PALAEOCENE	65 m.			
MESOZOIC		CRETACEOUS	136 m.		CHALK DEPOSITED. CLAYS AND SANDS.	SUB-TROPICAL
		JURASSIC	195 m.		LIMESTONES AND IRON-ORE DEPOSITED.	
		TRIASSIC	225 m.		SALT AND GYPSUM DEPOSITED.	DESERT
PALAEOZOIC	UPPER PALAEOZOIC	PERMIAN	280 m.		ROCK-SALT AND POTASH DEPOSITED.	
		CARBONIFEROUS	345 m.	VARISCAN	COAL MEASURES DEPOSITED. LIMESTONE DEPOSITED.	EQUATORIAL
		DEVONIAN	395 m.		SOUTHERN UPLANDS FAULT FORMED.	DESERT
	LOWER PALAEOZOIC	SILURIAN	440 m.	CALEDONIAN	HIGHLAND BOUNDARY FAULT. GREAT GLEN FAULT.	SUB-TROPICAL
		ORDOVICIAN	500 m.		MOINE THRUST INITIATED.	
		CAMBRIAN	570 m.			
PRE-CAMBRIAN						

FOSSILS

CORALS
BRACHIOPODS (rare)
GASTROPODS
TRILOBITES (rare)
LAMELLIBRANCHS
GRAPTOLITES
GRASSES
LAND PLANTS
AMMONITES
ECHINOIDS (rare)
MAMMALS (rare)

The major continental blocks on the Earth's surface are actually floating on a denser form of rock which is so hot and under such pressure that it actually flows and is "plastic" in its movement.

Geologists refer to the continental rock as SIAL, and to the plastic material below as SIMA.

Notice that where there is a mountain range there is a "root" below it, going deep into the SIMA on which the continent is floating. This is just the same as an iceberg where the portion visible above the water is supported by the rest floating in the water. The term isostasy is used to describe this and is applied to continents floating in SIMA.

ISOSTASY

The major continental blocks at the Earth's surface are actually floating on a denser form of rock which is so hot, and under such pressure that it actually flows and is "plastic" in its behaviour.

Geologists refer to the continental rock as SIAL, and to the plastic material below as SIMA.

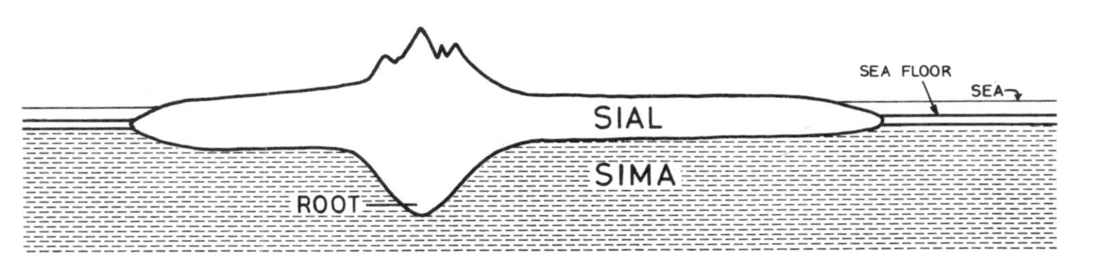

Notice that where there is a mountain range there is a "root" below to support it since the continent is floating. This is just the same as an iceberg, where the portion visible above the water is supported by the rest floating in the water. The term **Isostasy** is used tc describe the concept when applied to continents floating in SIMA.

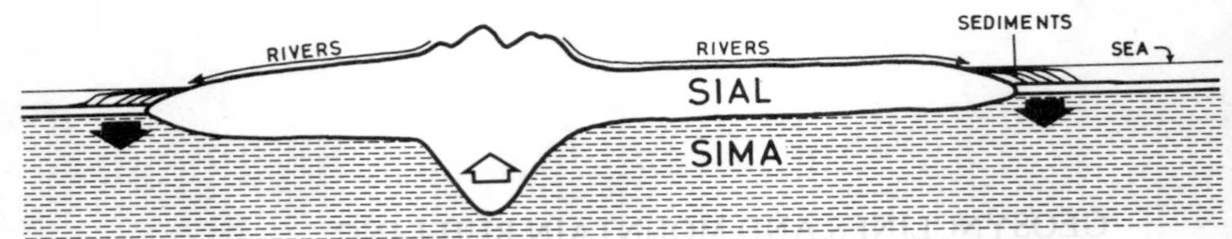

In the second diagram, it can be seen that erosion has started cutting the mountains down, and the rivers have deposited the sediments out to the sea. Since the mountain range is floating, and since the rivers have removed some material, the mountains slowly start rising in order to float a little higher, and re-establish a balance. This is known as **isostatic adjustment**. Similarly, in the coastal regions, the extra weight of the deposited sediments causes this part to start sinking. The great mountain ranges of the Alps and the Himalayas are rising slowly now as a result of isostatic adjustment. The Mississippi delta is sinking steadily under the weight of all the sediments being deposited in the sea by the Mississippi river alone.

GEOSYNCLINES AND MOUNTAIN BUILDING

The whole process of mountain-building can be divided into two basic stages.

(i) The geosynclinal stage. During this period the sediments that will ultimately become the mountain ranges are laid down and converted to rock.

(ii) The uplift stage, or mountain-building proper. During this period, the rocks are lifted out of the sea, and gently folded to form the mountain ranges.

GEOSYNCLINAL STAGE

The Earth loses heat to outer space by radiation. Its interior is very hot, but the continental blocks tend to act as "insulators" and confine maximum heat-loss to the oceanic areas. Consequently, the SIMA immediately beneath the continents tends to be hotter, while that under the oceans tends to be cooler.

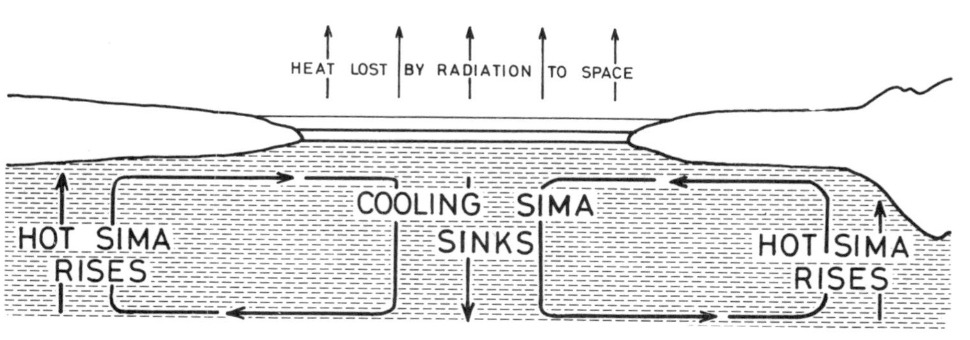

As the SIMA under the continental blocks becomes hotter, it becomes less dense, and starts to rise. Similarly, the cooler SIMA under the oceans starts to sink. A convection current is established between the two, with the hot SIMA spreading out from beneath the continents to replace the oceanic SIMA which is sinking.

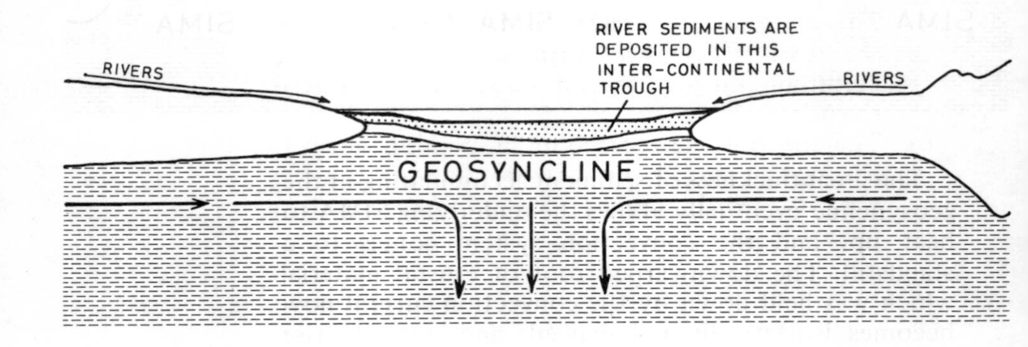

The first effect of this convection current is to downwarp the sea floor, and to keep it sinking. Consequently, sediments from the adjacent land masses pour in, and collect in great thickness. It should be noted that, in the case of a geosyncline, the sinking of the sea floor comes first, and it is the sinking that causes the great sediment accumulation. It is not an isostatic response to sediment deposition. The area of sedimentation consists of a series of individual elongate basins within a generally-subsiding through.

UPLIFT STAGE

The diagram below shows the effect of the SIMA convection current system after a long time.

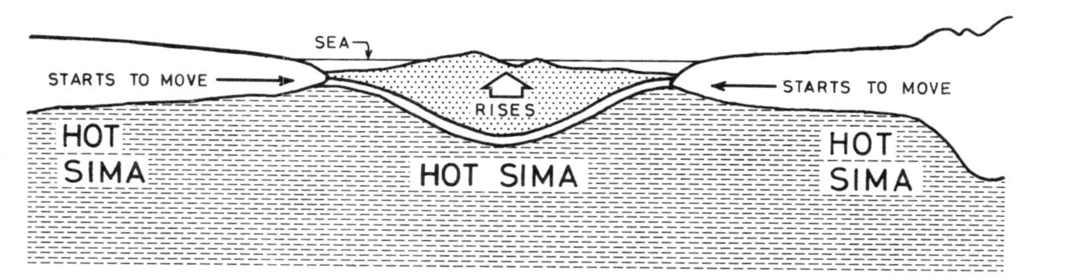

The great thickness of sediment accumulating in the geosyncline eventually allows the underlying SIMA to heat up, and the convection system comes to a standstill. The downwarped root of crustal material which was held down by the convection current now becomes isostatically unbalanced, and starts to rise rapidly in order to establish equilibrium.

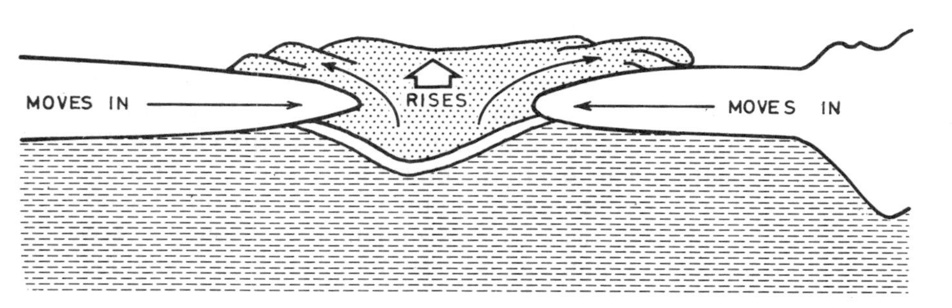

The friction of the convection currents may have imparted a motion to the overlying continents which now continue to move toward each other even though the convection currents have waned. The rising geosynclinal sediments are squeezed out over the continental blocks to form a new mountain range.

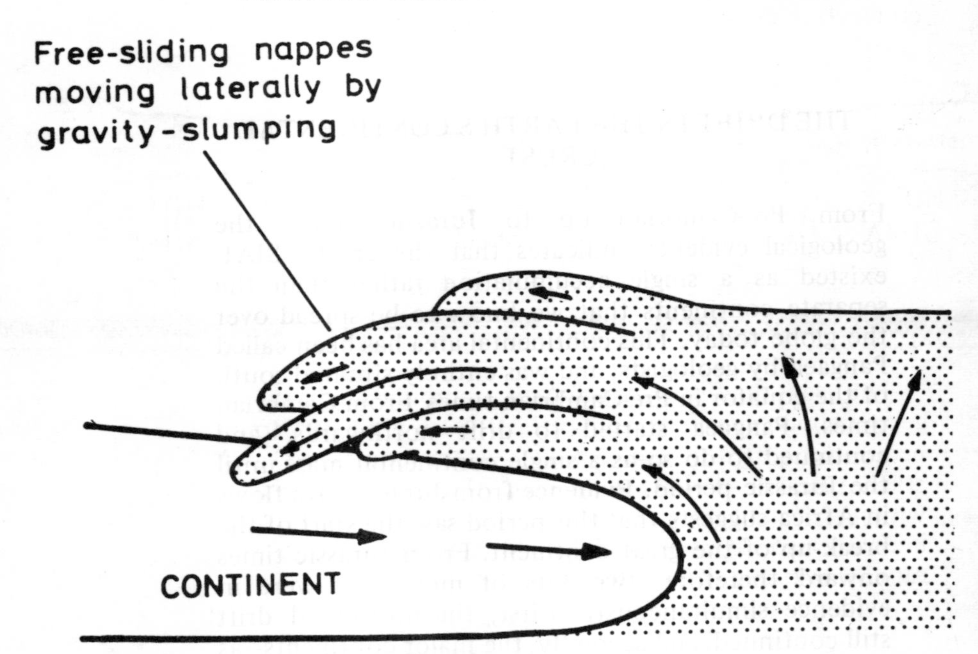

As the geosynclinal sediments rise, the marginal masses become unstable, and owing to the plasticity which the rocks exhibit at this stage, the overfolded masses begin to slide laterally in response to gravity. Softer clay and salt beds form slip planes over which the more rigid strata gravity glide, thus often travelling considerable distances over the moving continental foreland.

THE DRIFT IN THE EARTH'S CONTINENTAL CRUST

From Pre-Cambrian up to Jurassic times, the geological evidence indicates that the crustal SIAL existed as a single accumulation rather than the separate continents that we know to be spread over the globe today. This continental area has been called Pangaea by geologists, and was entirely situated south of the equator in Pre-Cambrian times. From Cambrian times, Pangaea started to drift northwards, and continued to do so as a single continental mass until the Jurassic period. Evidence from Jurassic lava flows in Africa suggests that this period saw the start of the break-up of the great continent. From Jurassic times onward therefore, two sets of movement were in evidence simultaneously. First, the northward drift still continued, and secondly, the major continents–as known now–spread away from one another. As far as we know, both processes are still in operation.

Figure A below shows the position of the major continent during Carboniferous times; the relative situations of the present-day continents is also shown. Notice how the American, European and Asian coal fields are situated, and how the Carboniferous glaciation known to have affected South America, South Africa, India and Australia, is explained by the Carboniferous reconstruction. Figure B shows the continents beginning to drift towards their present-day positions.

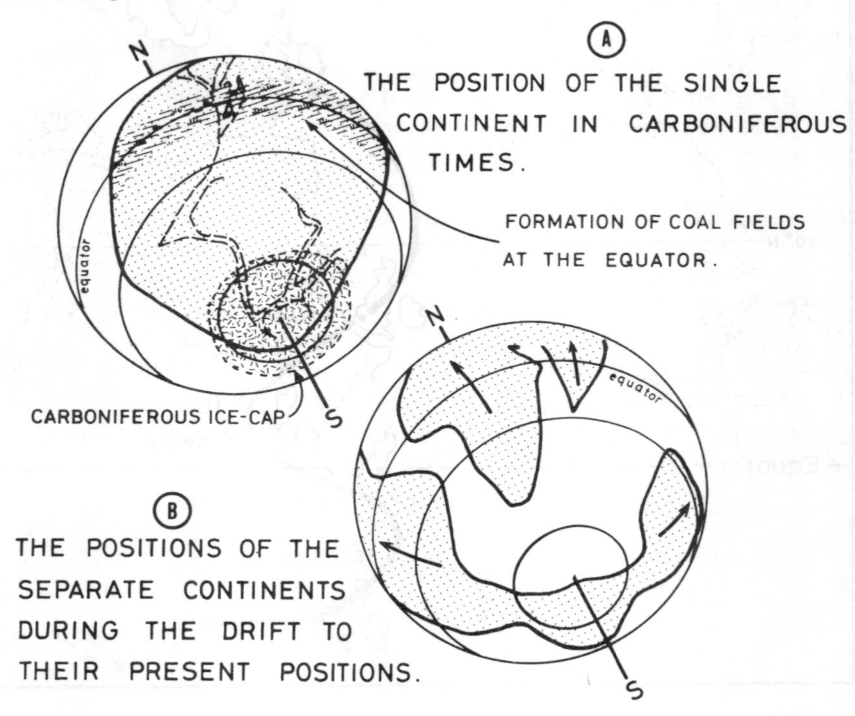

The most important evidence for Continental Drift comes from sedimentary and igneous rocks which display residual magnetism. The magnetism in these rocks is related to the position of the Earth's magnetic poles at the time of their formation. From a study of such rocks, the above conclusions can be drawn with regard to Continental Drift, and an estimate of the position of the British Isles in relation to its present position can be made as in the diagram below.

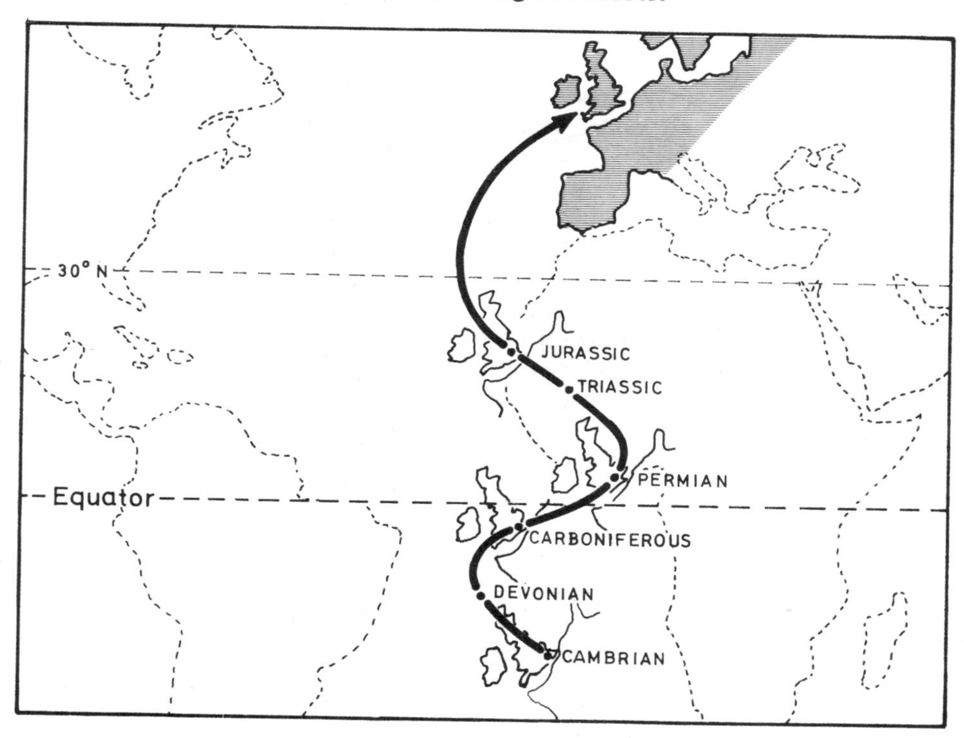

PRE-CAMBRIAN

While the Pre-Cambrian Rocks are estimated to range from 5,000 million up to 570 million years old, it is only in the most recent ones that fossil remains have been found. Pre-Cambrian fossils consist mainly of marine microfossils and the occasional sedimentary imprint of single soft-bodied creatures such as jelly-fish. Thus, although the rocks are of economic value in places on account of their mineral deposits, they are of little value to the stratigrapher. The few simple fossils, widely spaced, provide very little guide to the relative ages of Pre-Cambrian rock formations.

It is important that the reader should be fully aware of the immense stretch of time embraced by the term "Pre-Cambrian", in comparison with the relatively short 570 million years which are to be considered in further detail during the course of this book.

The old Pre-Cambrian rocks have been extensively folded and metamorphosed, and subsequently eroded to form subterranean basements for more-recent and less complex systems to rest upon. It is the different structural features caused by various periods of mountain building that are used to ascertain the relative ages of the major Pre-Cambrian rock groups. For instance, in Scotland, the depositional age of the oldest Pre-Cambrian rocks (The Lewisian) cannot be determined, and yet they must be at least 2,600 million years old since they were first affected by the Scourian Metamorphism at that time.

PRE-CAMBRIAN OUTCROPS OF THE BRITISH ISLES

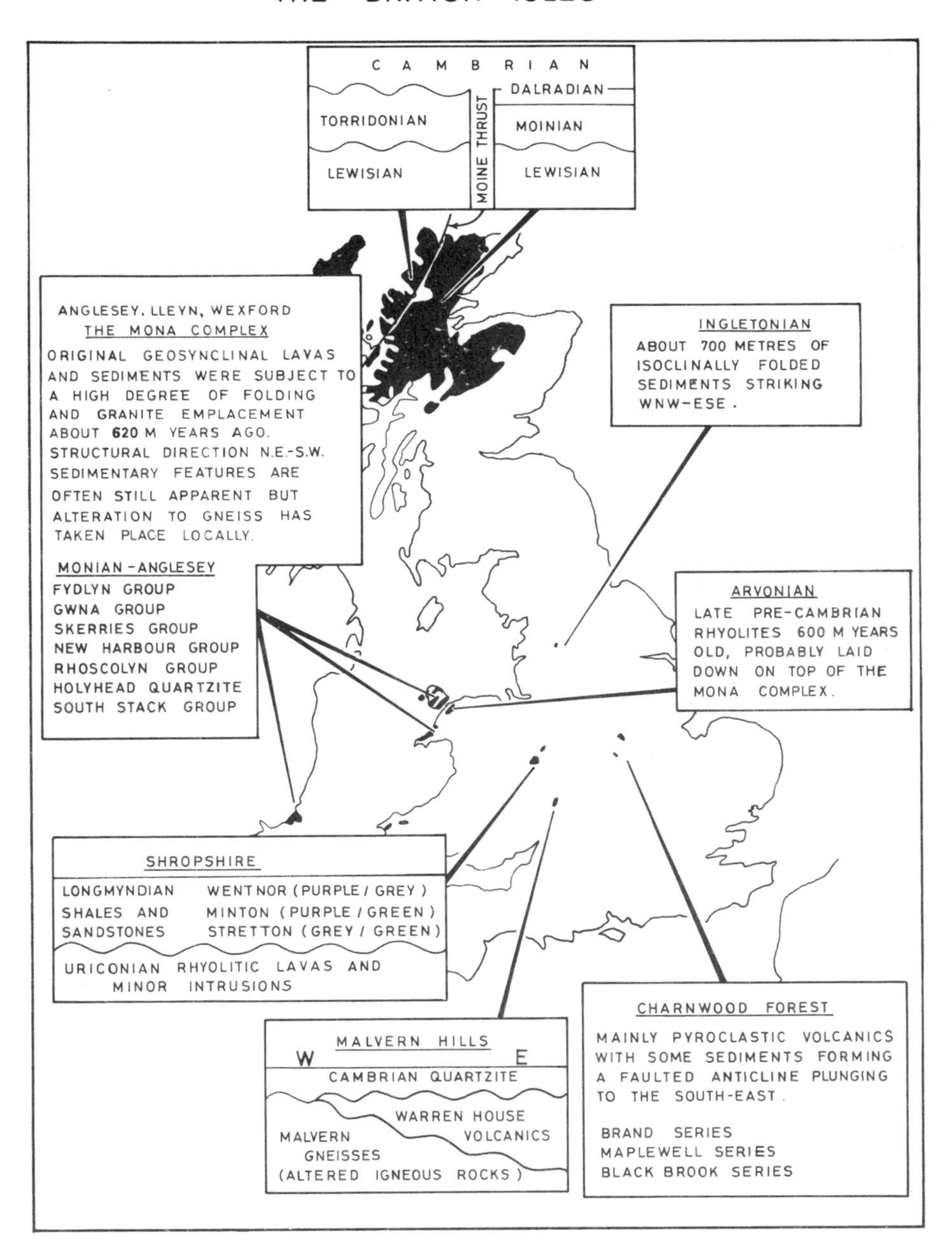

SIMPLIFIED BLOCK DIAGRAM OF THE PRE-CAMBRIAN BASEMENT ROCKS OF NORTH-WEST SCOTLAND

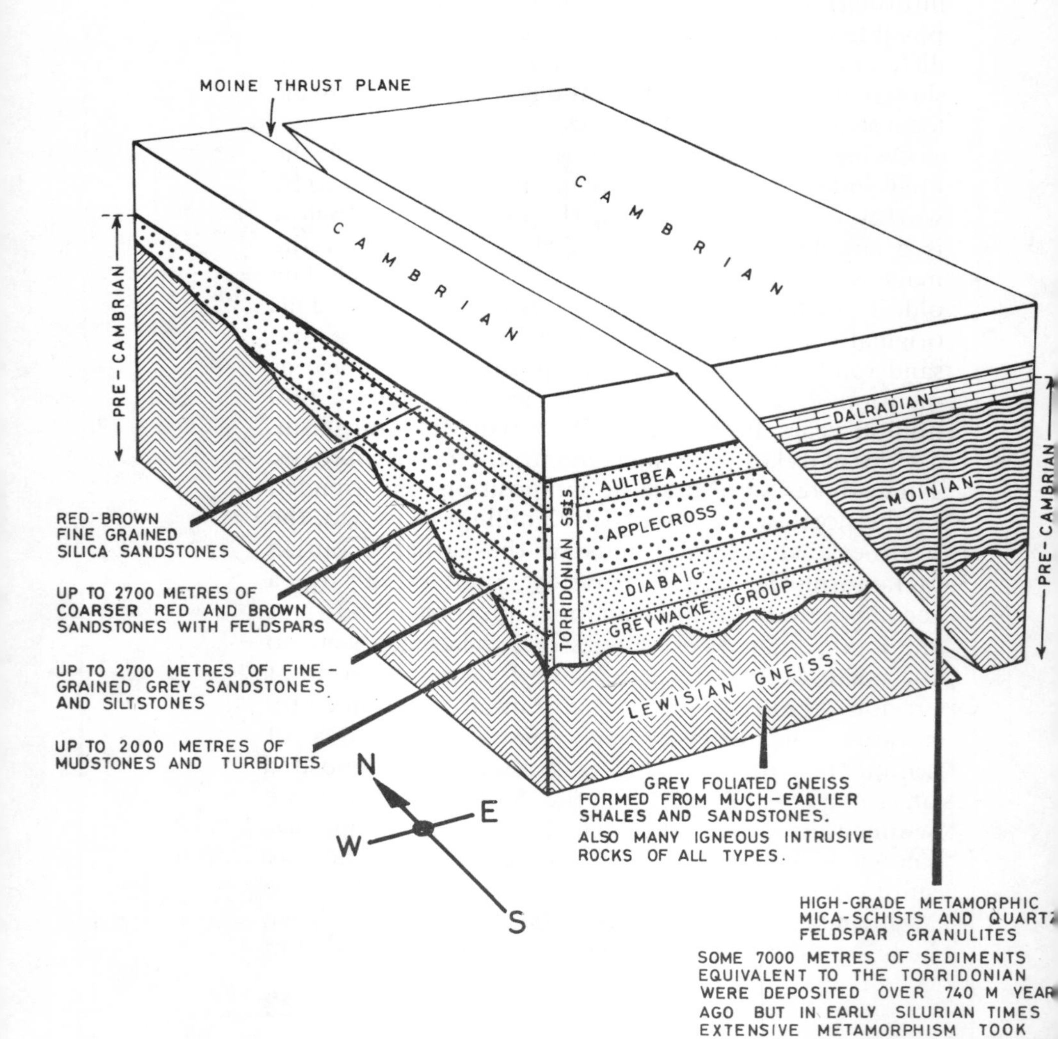

SOME 7000 METRES OF SEDIMENTS EQUIVALENT TO THE TORRIDONIAN WERE DEPOSITED OVER 740 M YEARS AGO BUT IN EARLY SILURIAN TIMES EXTENSIVE METAMORPHISM TOOK PLACE TO FORM THE MICA-SCHISTS AND GRANULITES.

A substantial part of the Scottish Highlands consists of Pre-Cambrian rocks outcropping at the surface, but in the rest of the British Isles, they form only very minor outcrops at isolated localities. From the limited information available from these outcrops, it is not possible to construct any palaeogeographical maps for different stages in the Pre-Cambrian, and so the map shown illustrates the present-day outcrops showing local sequences and rock types.

Owing to the large surface outcrop of the Pre-Cambrian rocks in Scotland, it is considered to be worthy of further mention: the block diagram below is a simplified illustration of the disposition of the main rock-types exposed in Northern Scotland. The oldest rock visible is the Lewisian Gneiss. This originally consisted of sedimentary shales and sandstones at the time of deposition. Their age as mentioned above, must be greater than 2,600 million years. Following the Scourian Metamorphism, a series of dykes were injected into the Lewisian rocks. These dykes were basic in composition, and trended to the north-west. About 1,600 million years ago, the north-east structural trends of the earlier Scourian Metamorphism were partially obliterated by a second period of folding and metamorphism known as the Laxfordian. These later structural trends were to the north-west. After this, there followed a period of erosion resulting in a substantial unconformity between the Lewisian and the 7,000 metres of sedimentary rocks which were subsequently deposited on top. To the west of the Moine Thrust these sedimentary rocks are virtually unaltered, and are known as the Torridonian Series. To the east, they are substantially metamorphosed and are known as the Moinian Series (or Moine Schists). The basal Cambrian beds are laid unconformably on the Torridonian, but this erosive stage is absent in the east, and the Moine Schists continue upward into the Dalradian Series which passes conformably up into the Cambrian Period. The evidence for this is that Lower Cambrian trilobites have been found in the upper Dalradian beds. The top of the Pre-Cambrian is therefore considered to be in the middle of the Dalradian Series.

CAMBRIAN

There are two main features which characterise the Cambrian all over the world. The first is the vast explosion of life which occured in the seas. The second is the world-wide submergence of vast areas of what had been old Pre-Cambrian land surface. (In Scotland for instance, the basal marine beds of the Cambrian can clearly be seen to overlie—unconformably—the Torridan Sandstones of the Pre-Cambrian.

These new Cambrian seas fostered the development of life, and the warm, shallow, spreading waters took the creatures with them and formed a unique marker horizon for the beginning of stratigraphical work.

The Cambrian period lasted about 70 million years, and the great submergence was not complete until nearly the end of this time. During this time, a wide variety of creatures had evolved and established themselves; they were predominantly "invertebrates", having no internal skeletons, but external shells to support and protect their bodies. These shells are easily preserved, and so make excellent fossils for study. Corals are unknown from the Cambrian, and the two main creatures used for stratigraphical studies are trilobites and brachiopods.

CAMBRIAN

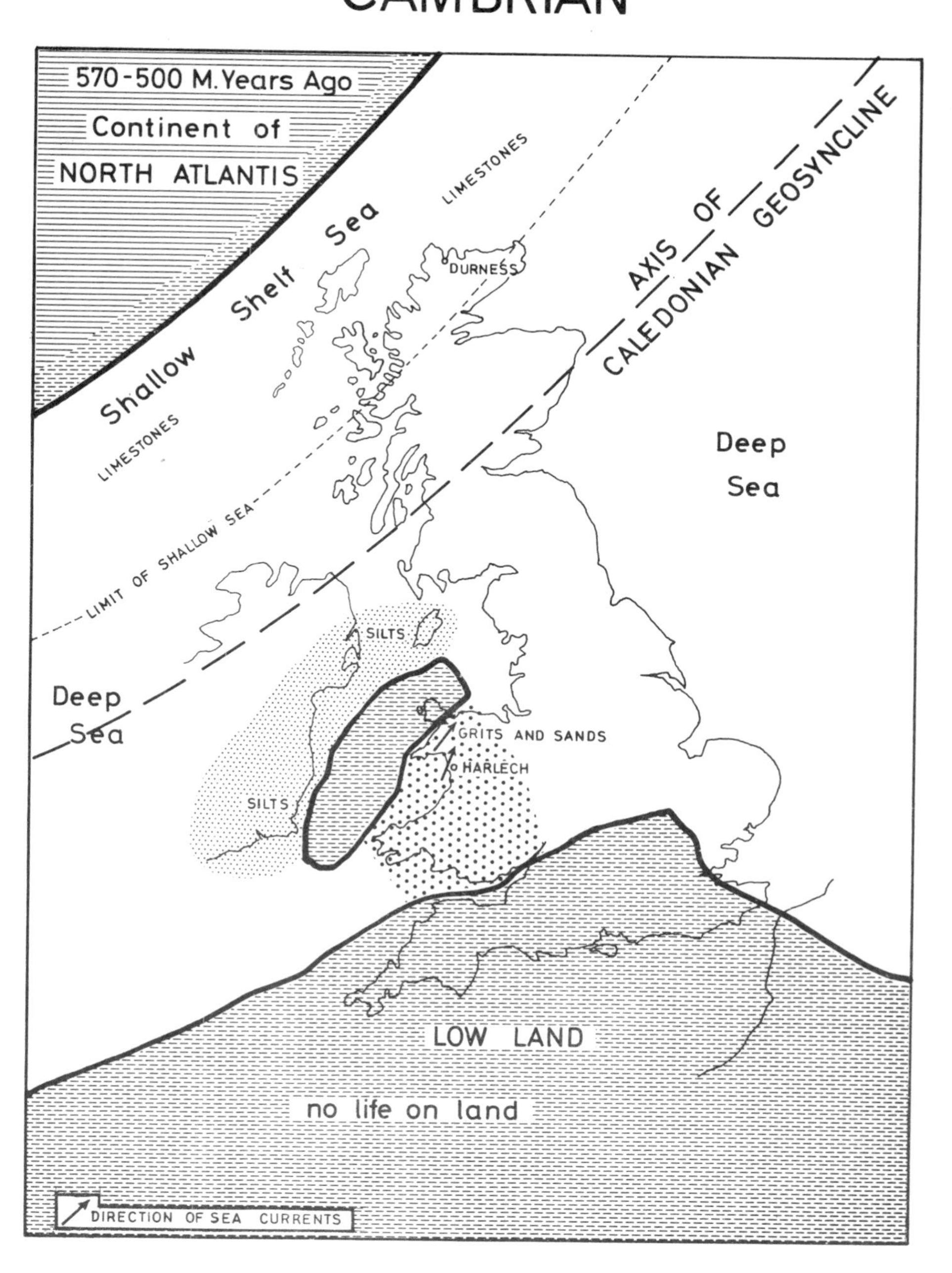
570-500 M. Years Ago
Continent of
NORTH ATLANTIS
Shallow Shelf Sea
LIMESTONES
LIMESTONES
DURNESS
AXIS OF CALEDONIAN GEOSYNCLINE
LIMIT OF SHALLOW SEA
Deep Sea
Deep Sea
SILTS
SILTS
GRITS AND SANDS
HARLECH
LOW LAND
no life on land
DIRECTION OF SEA CURRENTS

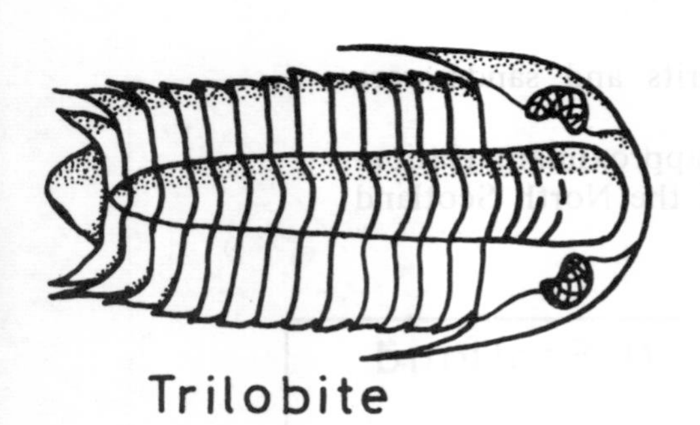

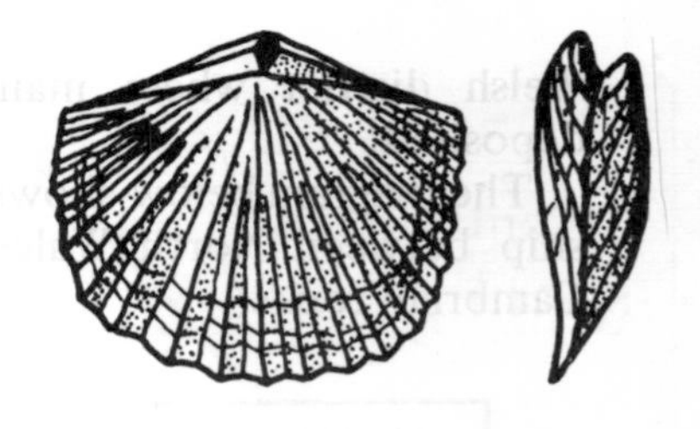

Trilobite Brachiopod

Over the area now occupied by the British Isles, a deep-sea zone existed, extending from the north-east to the south-west. The floor of the sea was rapidly sinking, and consequently great thicknesses of sediment were accumulating in this region. This is called the Caledonian Geosyncline. The sediments which are most characteristic of a geosynclinal environment are turbidites and greywackes. These rocks are sedimentary siltstones typically displaying graded-bedding; the lower portion of the beds contains coarser sediment than the upper. These are formed in the following way: disturbed sediments rush down the sloping sides of the geosyncline in the form of a suspension and spread out over the floor of the sea. From the suspended cloud of particles, the larger ones settle out first, followed later by the finer particles. Volcanic activity is sometimes connected with geosyncline formation, but in the case of the Cambrian, volcanic rocks are rare.

Against the coastline of the continent of North Atlantis, a continental-shelf zone spread out south-eastward, and did not form part of the sinking geosyncline. Much thinner rocks were forming here, and were in fact mainly dolomitic limestones, differing from the southern marginal zone in the

Welsh district where mainly grits and sands were deposited.

The diagram below shows the approximate relationship between North Wales and the North Scotland Cambrian rocks.

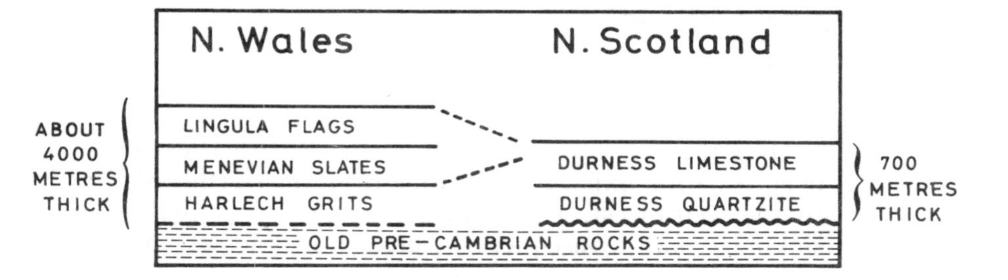

ORDOVICIAN

By middle Ordovician times, the large Caledonian Geosyncline had divided into two. The orogenic movements causing this division were the start of the Caledonian mountain building period which lasted about 100 million years and ended in Mid-Devonian times. The crustal pressures were from the south-east to the north-west and were great enough to form the Moine Thrust fault. Rocks of Pre-Cambrian age were thrust north-westward over Cambrian rocks formed on the old Cambrian sea shelf. The movement is estimated at some 16 km.

One of the main features of Ordovician times is the considerable amount of volcanic activity which took place around the margins of the geosynclines. A considerable part of the total thickness of the Ordovician rocks consists of rhyolitic and andesitic volcanic deposits, many of which consist of explosion debris, indicating the violence of the eruptions.

The seas in the region of the Lake District and Wales are shallow, with volcanoes standing up through the water forming archipelago chains of islands. The volcanic outpourings spread down beyond the island shores into the seas, thus passing laterally into sedimentary deposits. Typical of these is the Borrowdale Volcanic Series of the Lake District where, in the type-locality of Borrowdale, the tuffs, ashes, and pyroclastic rock show evidence of a shallow water and explosive volcanic origin.

ORDOVICIAN

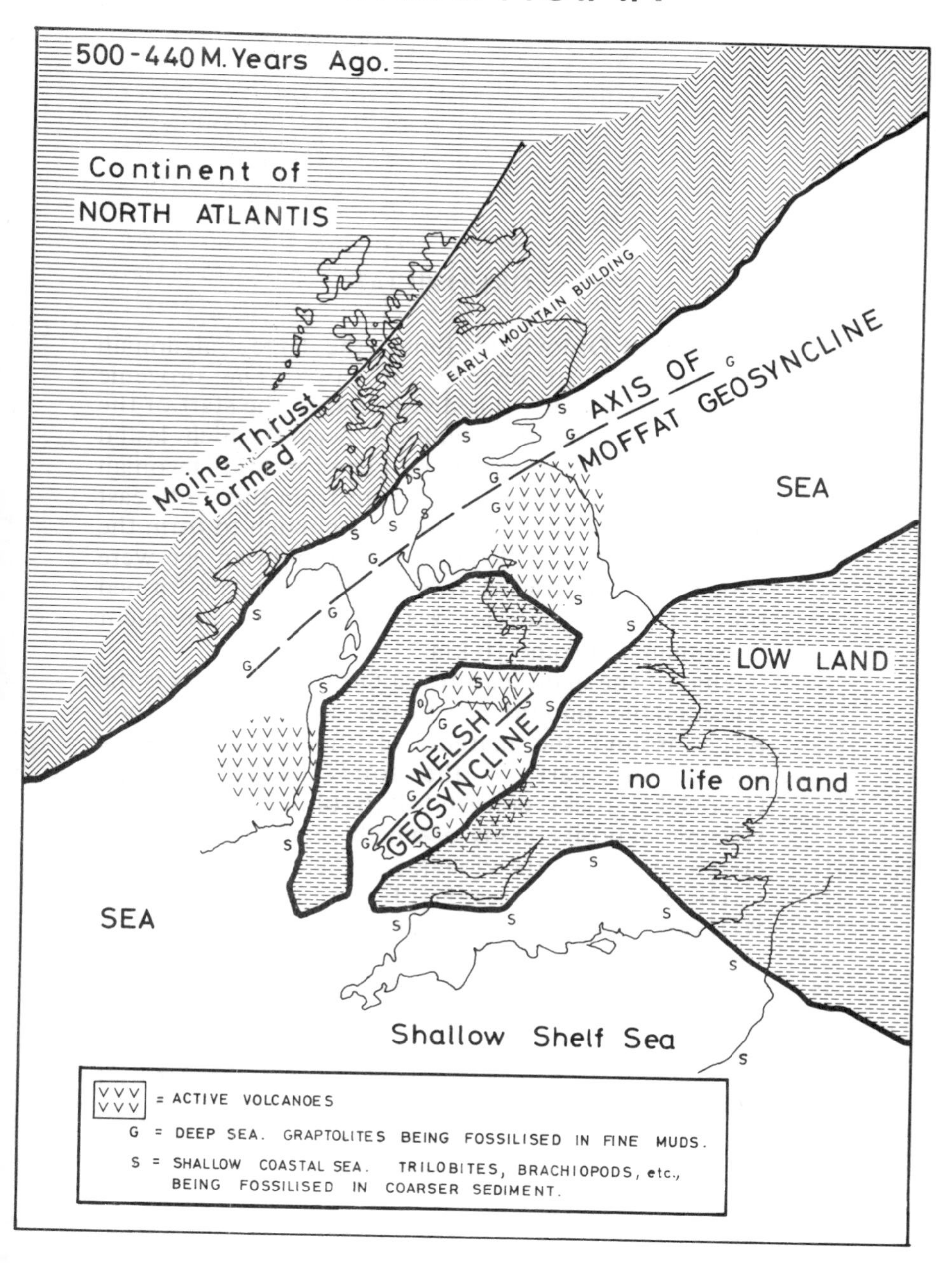
500-440 M. Years Ago.
Continent of
NORTH ATLANTIS
EARLY MOUNTAIN BUILDING
Moine Thrust
formed
AXIS OF
MOFFAT GEOSYNCLINE
SEA
LOW LAND
WELSH
GEOSYNCLINE
no life on land
SEA
Shallow Shelf Sea
= ACTIVE VOLCANOES
G = DEEP SEA. GRAPTOLITES BEING FOSSILISED IN FINE MUDS.
S = SHALLOW COASTAL SEA. TRILOBITES, BRACHIOPODS, etc., BEING FOSSILISED IN COARSER SEDIMENT.

In sedimentary rocks formed around the coastal zones, trilobites and brachiopods are plentiful, and are joined by the first corals. Also, at the beginning of Ordovician times, graptolites evolved. These creatures are important since their remains allow us to study the fine geosynclinal sediments in great detail. Graptolites evolved very rapidly, and therefore permit very fine sub-divisions to be established; there are some fourteen graptolite zones in the Ordovician. The reason for their wide distribution is that these creatures floated on the surface of the sea, and drifted with the ocean currents which swept them far beyond their normal environments. When they died their remains would sink to the sea floor, and thus unlike many other fossils, they are found contemperaneously in rocks indicative of substantially varying environment.

The diagram below shows the living mode of graptolites.

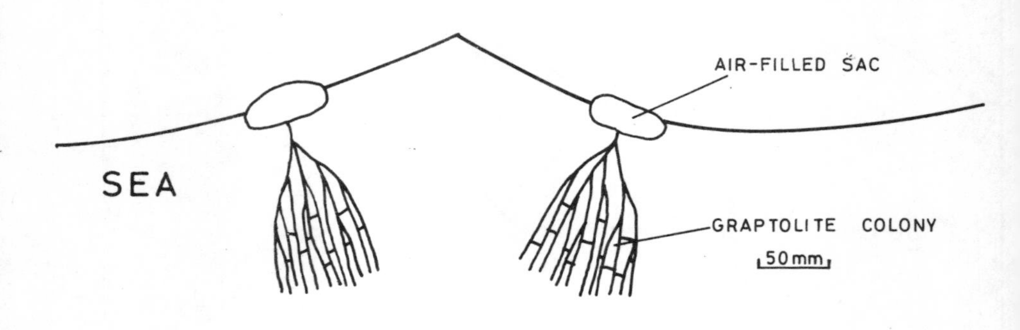

The trend in graptolite evolution was from more complex types toward more simple types.

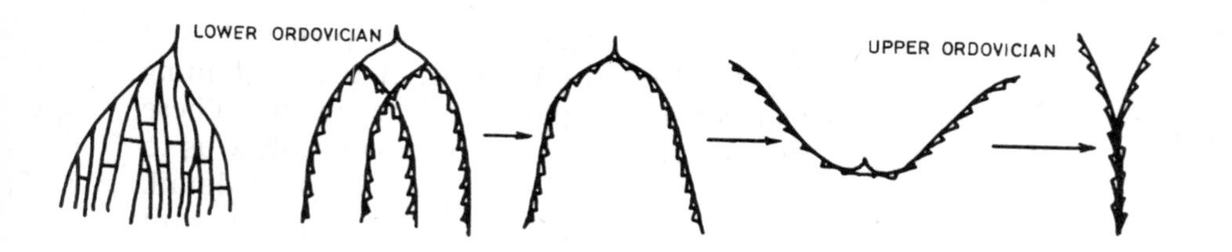

The following is an approximate correlation between the rocks of the North Wales area and those of the Lake District and Scotland.

	WELSH AREA		LAKE DISTRICT		S. UPLANDS	
U.ORD	ASHGILLIAN		ASHGILL SHALES		ASHGILLIAN	
	CARADOC	V V V	CONISTON LIMESTONE		CARADOCIAN	
L.ORD	LLANDEILO		BORROWDALE VOLCANIC SERIES	V V V V V V V V V	absent	
	LLANVIRN	V V V V V V				
	ARENIG	V V V	SKIDDAW SLATES		ARENIG	V V V
	TREMADOC				absent	

SILURIAN

For most of the Silurian, the geography was very similar to the Ordovician period, with the Welsh and the Moffat Geosynclines both active. It was not until late Silurian times that a surge in the Caledonian mountain building caused a major change in the geography. It is this late Silurian geography which is depicted in the adjacent map.

The Silurian period did not experience the marked volcanic activity that so characterised the Ordovician.

In early Silurian times, the Midland land-mass became submerged to form a shallow shelf-sea zone adjacent to the Welsh Geosyncline. In late Silurian, the sediments which had formed in the Moffat Geosyncline were pushed up to form the great Caledonian Mountain ranges, the structural roots of which still form the Scottish Highlands, and further north, the Norwegian mountain ranges, with typical south-west to north-east trending axes.

During this active mountain building period the Great Glen Fault was formed which still forms a major geographical feature today, and in which Loch Ness lies. Also during this period the Highland boundary fault was initiated, which today forms the northern edge of the Midland Valley region of Scotland.

While the Moffat Geosyncline entered into its mountain-building phase during the Silurian, the Welsh Geosyncline lasted right through to the end of

SILURIAN

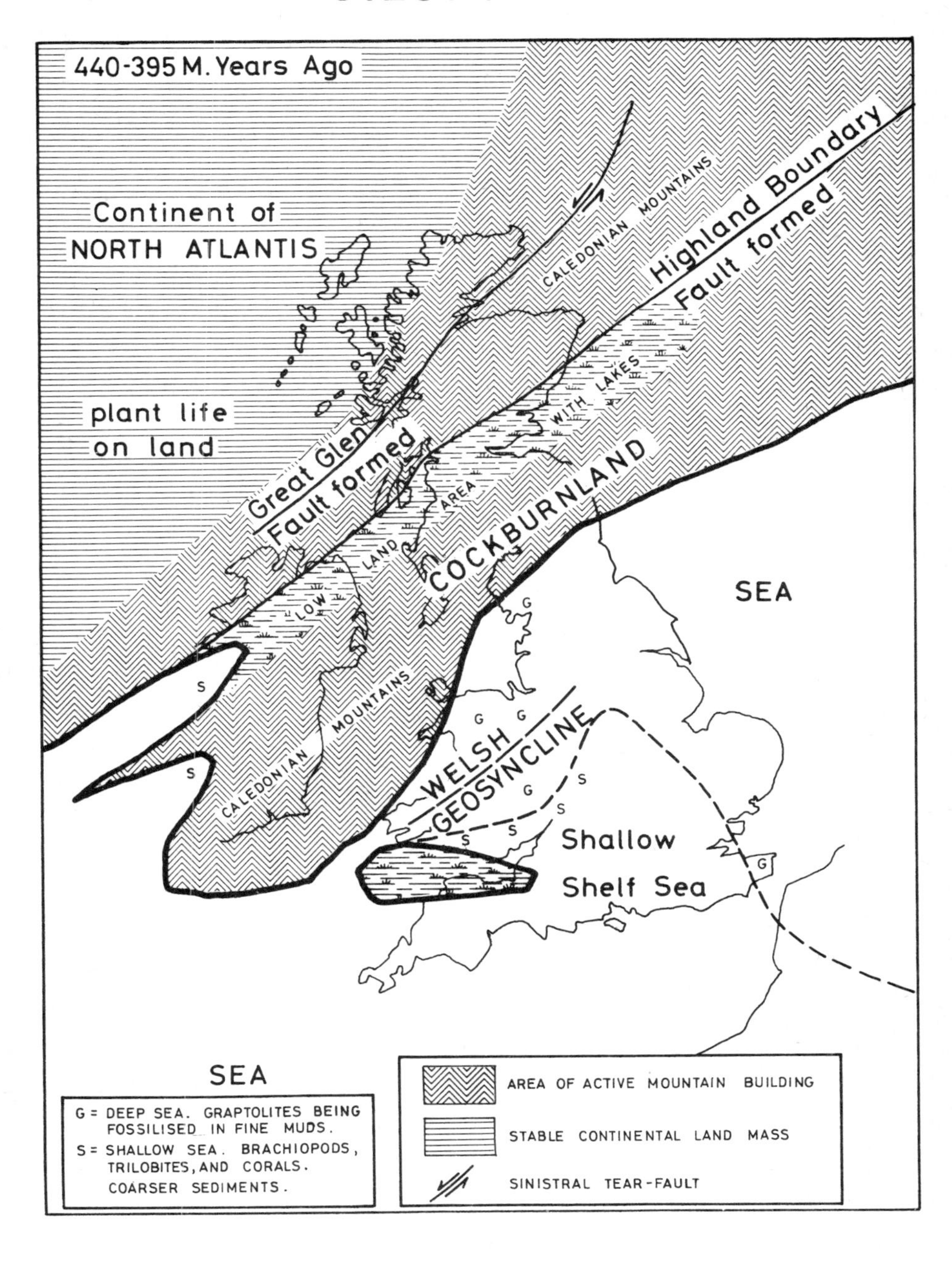

440-395 M. Years Ago
Continent of
NORTH ATLANTIS
plant life
on land
CALEDONIAN MOUNTAINS
Highland Boundary
Fault formed
Great Glen
Fault formed
LOW LAND AREA WITH LAKES
COCKBURNLAND
CALEDONIAN MOUNTAINS
SEA
WELSH
GEOSYNCLINE
Shallow
Shelf Sea
SEA
G = DEEP SEA. GRAPTOLITES BEING FOSSILISED IN FINE MUDS.
S = SHALLOW SEA. BRACHIOPODS, TRILOBITES, AND CORALS. COARSER SEDIMENTS.
AREA OF ACTIVE MOUNTAIN BUILDING
STABLE CONTINENTAL LAND MASS
SINISTRAL TEAR-FAULT

Silurian period. Consequently, in Wales, there is a record of all Silurian rocks, whereas for instance in the Southern Uplands area, Upper Silurian rocks are naturally absent since it was an area of uplift and erosion. The table below shows the basic subdivisions of the Silurian in Wales.

Trilobites and brachiopods are still the most useful fossils from the shallow coastal sediments, but are joined by the abundant remains of corals which formed reefs. Two other important "starters" at this time were early fish, whose remains are particularly found in the lowland area lakes south of the Highland Boundary Fault, and very important, the first life on land—plantlife—is recorded from the Silurian.

The graptolites continued to be fossilised in the deep-water areas, and provide an excellent record to the top of the Silurian where they die out. They were mainly single-stemmed graptolites of the Monograptus type.

LUDLOW	MUDSTONES WITH NODULAR LIMESTONES
WENLOCK	SHALES AND LIMESTONES
LLANDOVERY	SANDSTONES AND SHALES

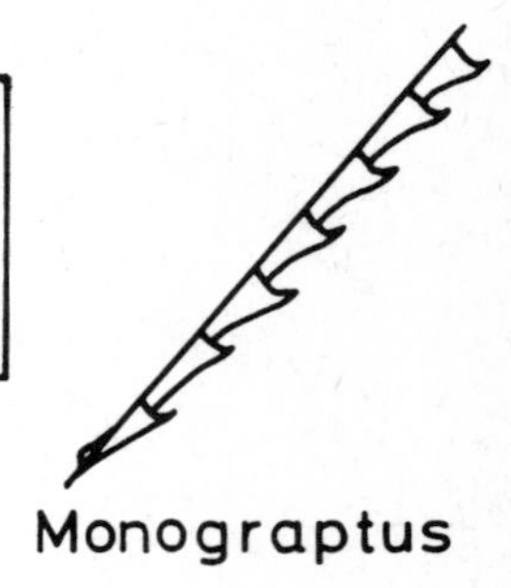

Monograptus

DEVONIAN

The Devonian marked the end of the Caledonian mountain-building period. In the final phases of the mountain-building nearly all the crust now occupied by the British Isles was raised above sea level. From the accompanying map it can be seen that only the south of England was experiencing marine conditions. Thus north of this zone continental rocks were being laid down in localised areas, whilst in others massive erosion was taking place. The desert sandstones and freshwater lake deposits of this period are often known collectively as "Old Red Sandstone".

In Scotland, the southern Uplands Fault was formed, thus completing the formation of the Midland Valley Graben, which has ever since acted as a negative area, and still today forms the Midland Valley of Scotland. In Devonian times, it was a low-lying area continually flooded with giant freshwater lakes in which lived the newly-evolved fish. Although the Great Glen Fault had been initiated in Silurian times, the main movement on the fault took place in early Devonian times.

In the south, the geosynclinal stage of the next mountain-building period had already started simultaneously with the close of the Caledonian mountain-building phase. Between the continental and geosynclinal facies was a shallow-water marine facies in which coarse sedimentaries, shelf, and reef limestones were forming. Typical marine fossils of this

DEVONIAN

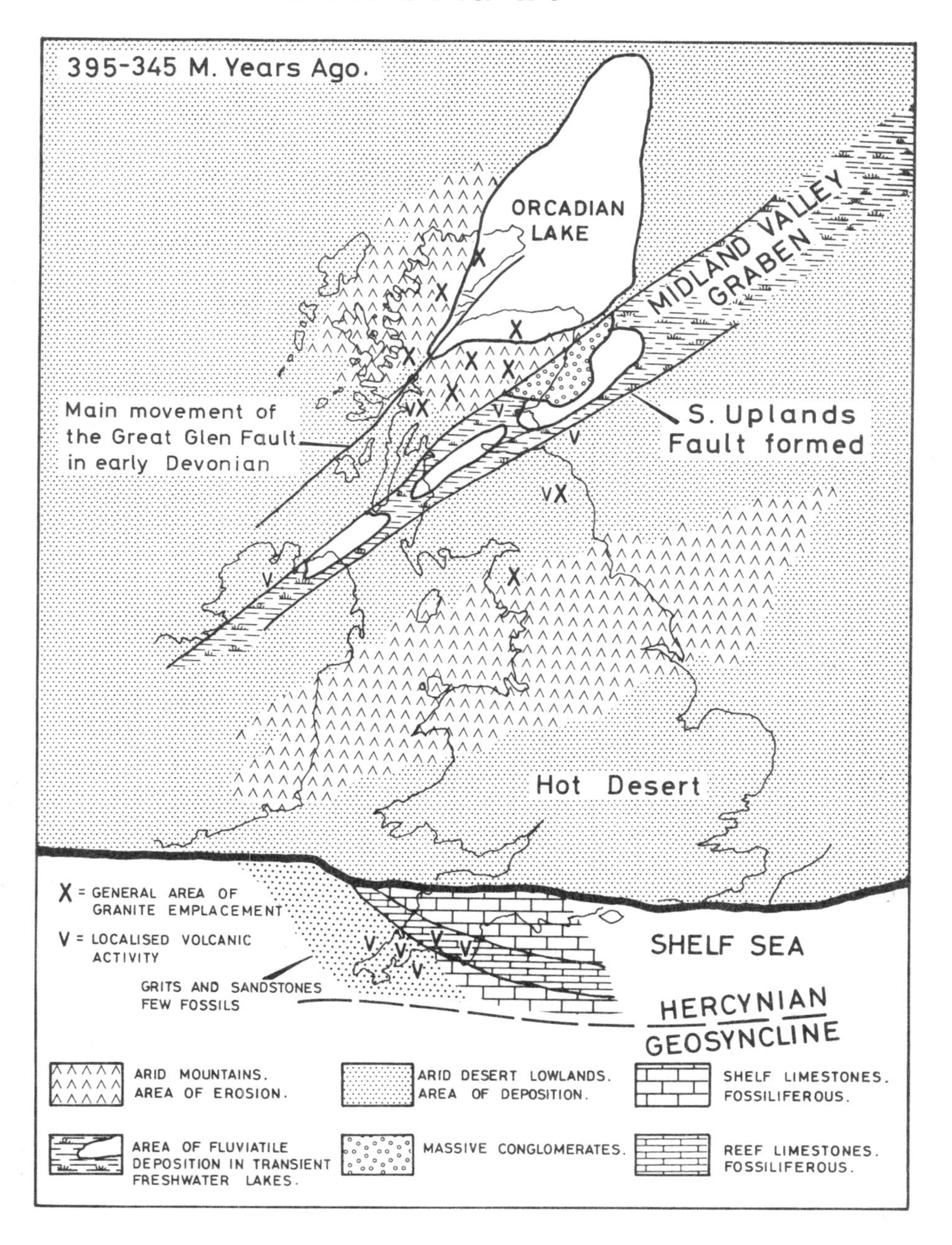

395-345 M. Years Ago.
ORCADIAN LAKE
MIDLAND VALLEY GRABEN
Main movement of the Great Glen Fault in early Devonian
S. Uplands Fault formed
Hot Desert
X = GENERAL AREA OF GRANITE EMPLACEMENT
V = LOCALISED VOLCANIC ACTIVITY
GRITS AND SANDSTONES FEW FOSSILS
SHELF SEA
HERCYNIAN GEOSYNCLINE
ARID MOUNTAINS. AREA OF EROSION.
ARID DESERT LOWLANDS. AREA OF DEPOSITION.
SHELF LIMESTONES. FOSSILIFEROUS.
AREA OF FLUVIATILE DEPOSITION IN TRANSIENT FRESHWATER LAKES.
MASSIVE CONGLOMERATES.
REEF LIMESTONES. FOSSILIFEROUS.

time were trilobites, gastropods, brachiopods, corals and newly-evolved goniatites.

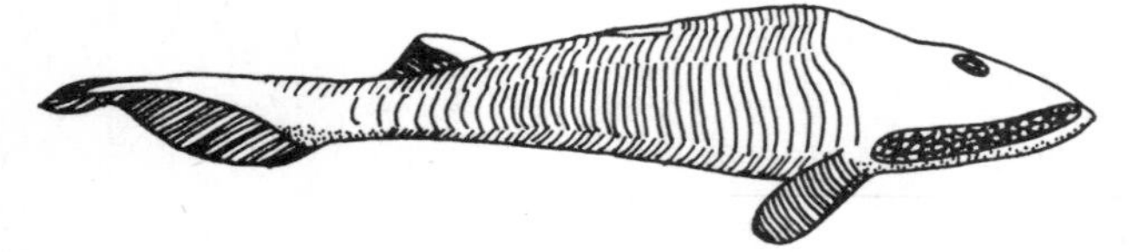

Early type of fish

Goniatite
(Early ammonoid)

In the continental rocks, however, correlation of zones is only possible by use of the newly-evolved fish faunas of the freshwater lakes. Since lakes are localised phenomena, the correlations are often very difficult. Land plants continued to evolve during the Devonian period, but were restricted in area by the prevailing desert climate.

Much vulcanicity took place in the Midland Valley Graben and on the marine shelf zone north of the Cornwall Geosyncline. In the Midland Valley the igneous rocks are mainly basalts and andesites and the extrusions more in the form of lava-flows than paroxysmal eruptions. Many granites were emplaced in Scotland (e.g. Ben Nevis), the Lake District (e.g. Shap) and the Cheviots.

The Devonian continental feature is one of tremendous denudation, and during this period the great Caledonian mountain systems were planed down to their roots. Thus the pre-Carboniferous landscape must have been a gently undulating surface across which the early-Carboniferous sea was to transgress.

	S. DEVON	N. DEVON	PEMBROKE. W. MIDLANDS	MIDLAND VALLEY	ORCADIAN LAKE
U.	OSTRACOD SLATES TORQUAY LIMESTONE	PILTON BEDS PICKWELL DOWN SANDSTONE	FARLOVIAN	UPPER O.R.S.	UPPER O.R.S.
M.	LIMESTONE	ILFRACOMBE BEDS	ABSENT	ABSENT	MIDDLE O.R.S.
L.	STADDON GRITS MEADFOOT BEDS	HANGMAN GRITS	BRECONIAN		ABSENT
	DARTMOUTH SLATES	LYNTON BEDS FORELAND GRITS	DITTONIAN	VVVVVVVVV VVVVVVVVV	ABSENT
			DOWNTONIAN		ABSENT

MARINE LIMESTONES SANDSTONES, SHALES AND SUBMARINE LAVAS.

TERRESTRIAL DESERT SANDSTONES, AND LAKE DEPOSITS INCLUDING SANDSTONES, MARLS, SHALES, AND CONGLOMERATES.

ZONE OF COASTLINE OSCILLATION.

CARBONIFEROUS

To summarise the Carboniferous, would be to say that the sea spread over the country from south to north, and was itself slowly filled-in by an enormous river delta which built out from the north-east to the south-west. It would have to be mentioned that right at the end of the Carboniferous period, the major part of the Hercynian Mountain Building activity took place. Consequently, it has been considered necessary to prepare a general map for the Carboniferous, from 345 – 280 million years ago, and a separate one to illustrate the effects of the Hercynian Mountain Building, during the later Carboniferous times from 290 – 280 million years ago.

The general Carboniferous map shows the country already covered by the Lower Carboniferous sea, and indicates the successive stages of advance of the delta building out from the north-east.

As the sea spread across the old Devonian land surface it generally formed a basal conglomerate on top of which were formed limestones; small shelly creatures thrived in the clear warm seas. By the end of Lower Carboniferous times an arm of the delta had spread as far as the area now occupied by the Pennines, and was laying down the gritstone deposits of this region. As the delta advanced further westward, it deposited the gritstone series often known collectively as the Millstone Grit.

CARBONIFEROUS

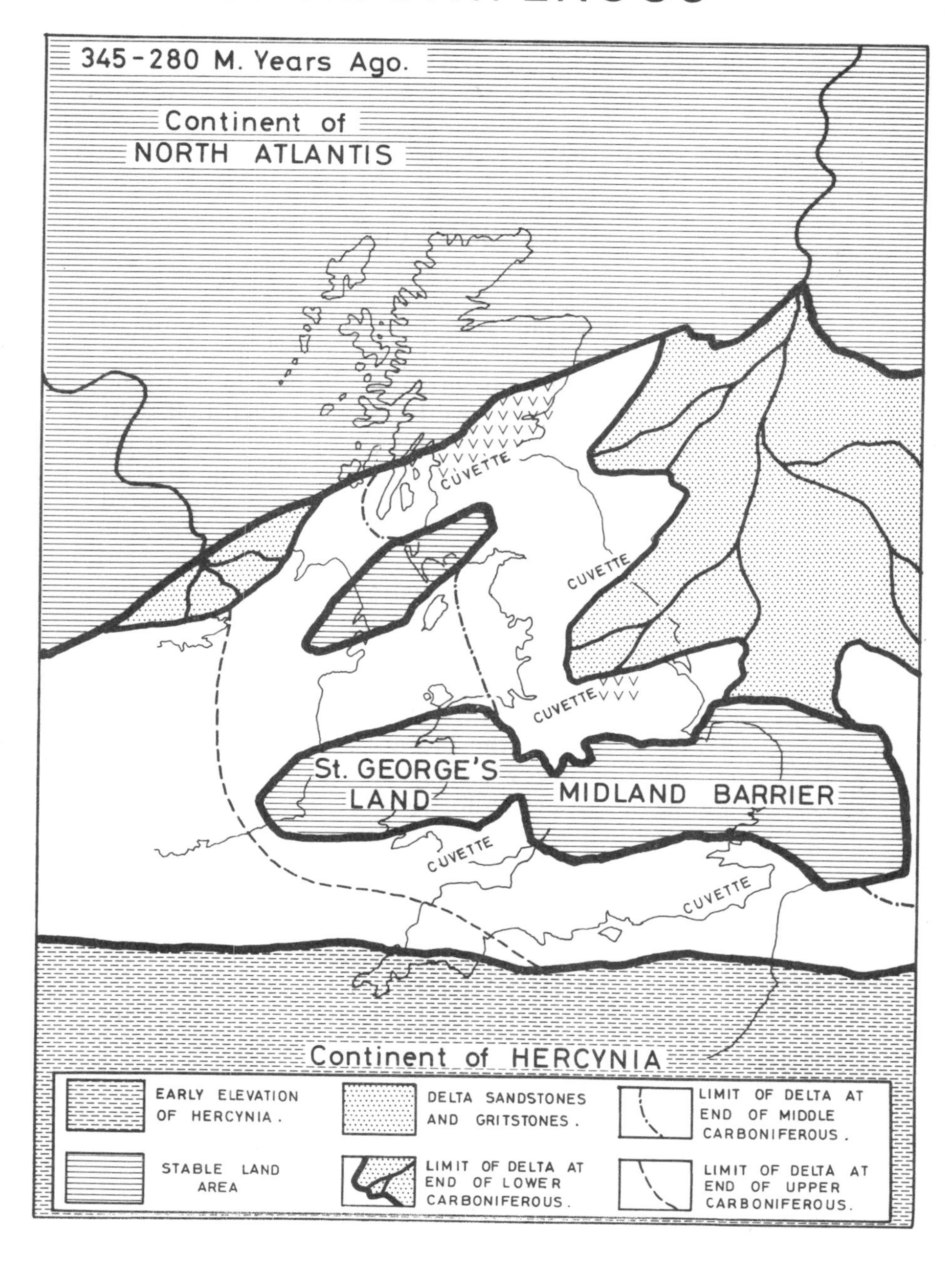

Localised areas of the delta were sinking more rapidly than others. The sinking was possibly due to both isostatic response to the weight of the sediments, and also to crustal instability cause by the impending Hercynian Mountain Building. These localised areas of sinking are referred to as "Cuvettes". They are not trough-like structures, nor are they as large as geosynclines, but are basin-like, and the subsidence both permits and causes the local accumulation of great thicknesses of sediment. The centre of a cuvette basin sinks more rapidly than the periphery, with a result that beds deposited in a cuvette are thicker towards the centre, and become thinner laterally towards the basin periphery. The dominant cuvettes of the time are illustrated on the map, and were active both during the Middle-Carboniferous, when the Millstone Grit was being accumulated, and also during Upper-Carboniferous times when the Coal Measures were forming.

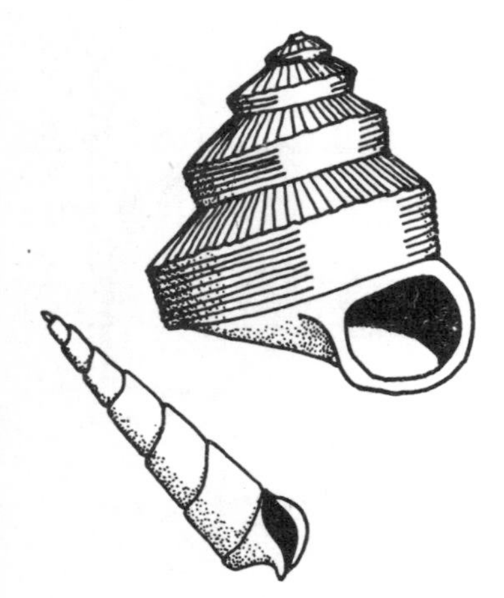

Gastropods.

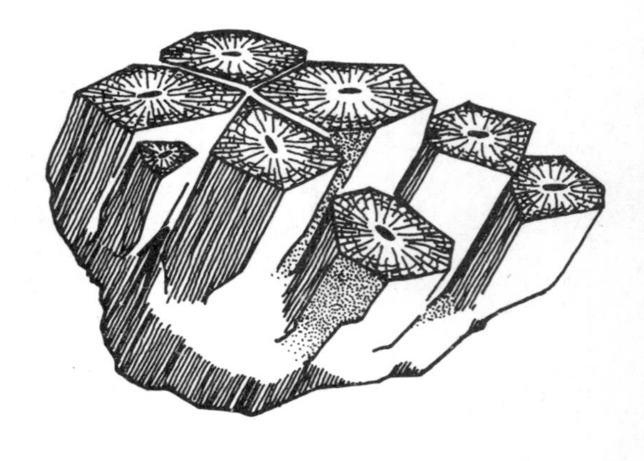

Colonial Coral.

There was localised volcanic activity in the Midland Valley of Scotland, and also in the region of the Pennines where lava flows can be found in the Lower Carboniferous limestones.

The coal seams of the "Coal Measures" are the accumulation of fossil vegetation from the equatorial forests which established themselves on the delta as it filled in the sea. Land plants had only just evolved, and since the equator was lying across the present British Isles at that time, the world's first major coal fields were established in this region. Insects had established themselves on land in the Carboniferous forests, and the first reptiles moved up onto the land surface in the moist swamps of the delta.

UPPER CARBONIFEROUS	STEPHANIAN	HERCYNIAN MOUNTAIN BUILDING
	WESTPHALIAN	EQUATORIAL SWAMPS FORMING THE COAL MEASURES
	NAMURIAN	DEPOSITION OF DELTA GRITSTONES
LOWER CARBONIFEROUS (DINANTIAN)	VISEAN	FORMATION OF LIMESTONES IN CLEAR, WARM SEAS
	TOURNASIAN	

HERCYNIAN MOUNTAIN BUILDING

(Stephanian. Upper Carboniferous.
290–280 M. years ago.)

The Stephanian is known for its mountain building activity rather than for any rock types deposited. In fact the earlier cuvette areas did continue to subside a little with meagre coal formation, but everywhere else the delta turned to red barren sand heralding the onset of desert conditions.

The general theme of the mountain building was that a crustal pressure built up from the south, causing severe folding and mountain building in the newly-formed sediments of the Hercynian Geosyncline of Devonian times. These mountain chains were built up with structural trends from east to west, with major northward thrusts in both Cornwall and France, and the emplacement of the Cornwall and Devon granites.

The major mountian front was limited to the extent of the Devonian marine rocks, and in the continental area to the north, the pressure caused largescale, but gentle, folding such as the uplift of the Malverns, and the Pennine range, and also some faulting of the Palaeozoic rocks. In the Midland Valley and the north of England two major sills were intruded at the end of the Stephanian. The Whin Sill in the north of England is the largest "minor-intrusive" body in the country.

THE MOUNTAIN-BUILDING PHASE OF THE HERCYNIAN (OR VARISCAN) OROGENY

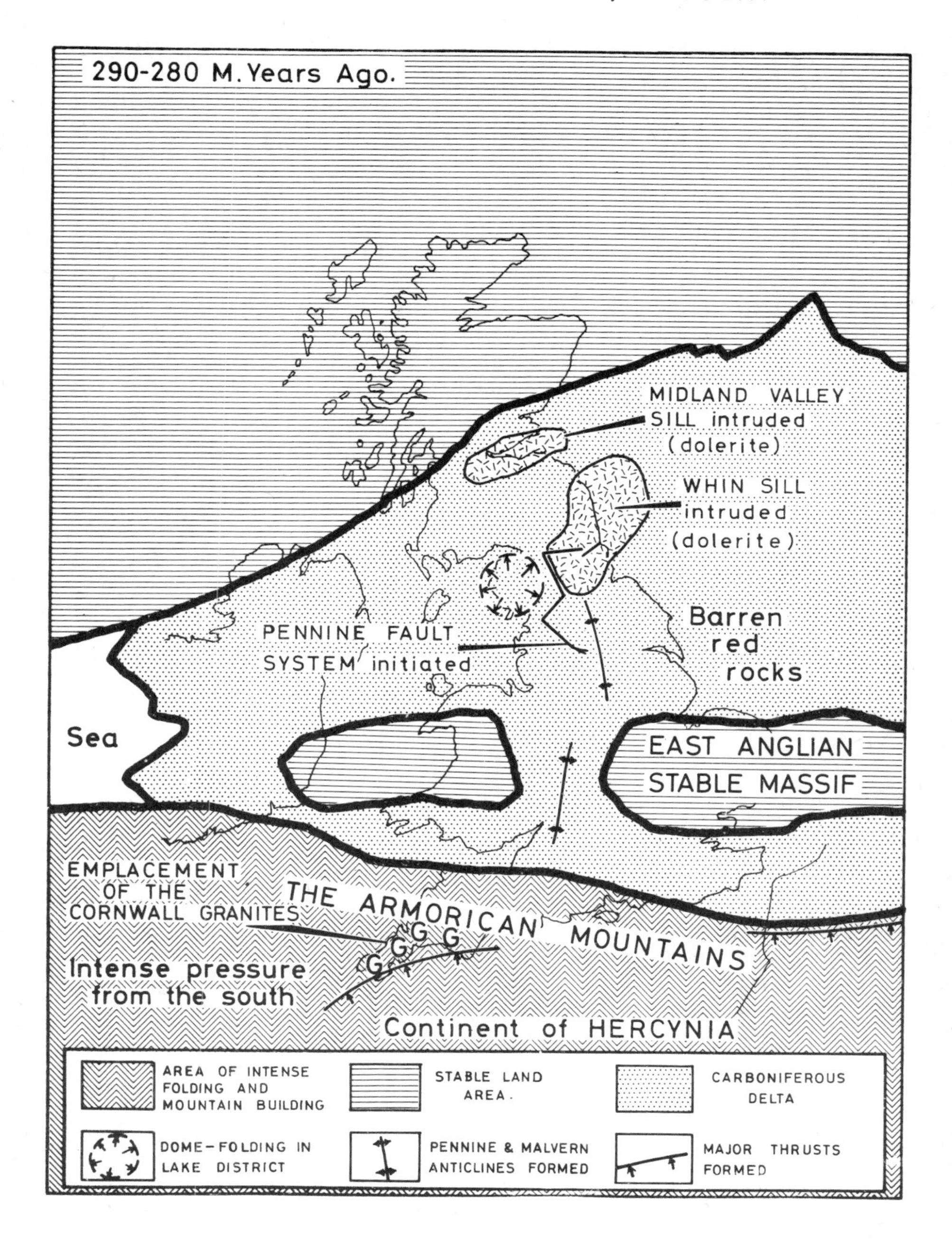

It should be noted that the fortuitous occurance of a mountain building period immediately after the formation of the coal fields effectively preserved many of the coal fields by down-folding. Coal is not a rock which resists erosion, and it is interesting to speculate as to how much of the Coal Measures would have been left had the Hercynian Mountain Building taken place earlier or much later. The positions of the present coal fields are therefore a combined result of localised accumulation and localised down-folding. The localised accumulation means that beds of coal thicken towards the centre of the ancient cuvette areas, becoming more economic to work at the centres, and less so at the edges where they are thin. The Stephanian mountain-building has also affected the working of the coal seams in that some areas have been steeply tilted and faulted (e.g. the Lancashire coal field), whilst other areas have not (e.g. the Nottinghamshire coal field).

PERMIAN

The dry climate indicated in the Stephanian rocks continued to become more marked, and by Permian times, hot desert conditions had set in. North Atlantis and the continent of Hercynia had joined together as a result of the Variscan Orogeny which ended in early Permian times.

By early Permian times the old Carboniferous landscape had been moulded into four subsiding cuvette areas surrounded by desert mountain regions. Occasional torrential mountain storms swept rubble down onto the desert plains to form breccias. Sand dunes travelled across the desert lowlands driven by a steady wind from the east and accumulated to form the lowest of the Permian sub-divisions–the Yellow Sands.

Following this, a large shallow sea formed across Europe, and the western arm of this sea flooded the low-lying cuvette of north-east England. The lowest bed of this marine incursion is only five metres thick but is remarkably persistent. Despite its name of "The Marl Slate", it is actually a shale. The beds which lie on top of the Marl Slate are the Magnesian Limestone Series. The lower and middle Magnesian Limestone beds are found in the north-east cuvette only, but the sea must have made a sudden sweep across the country in late Permian times since the upper Magnesian Limestone is found as far east as northern Ireland.

PERMIAN

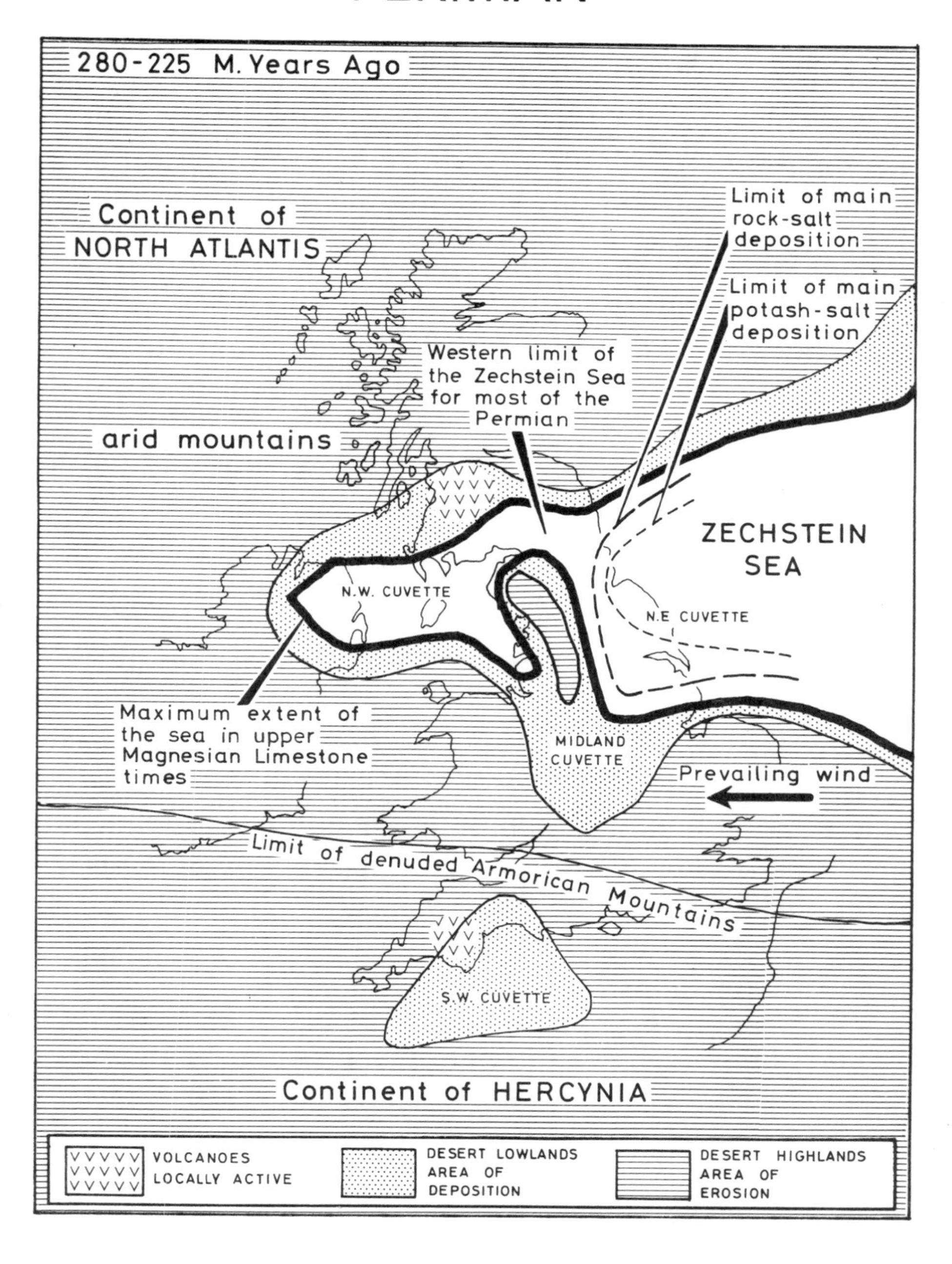
280-225 M. Years Ago
Continent of NORTH ATLANTIS
Limit of main rock-salt deposition
Limit of main potash-salt deposition
Western limit of the Zechstein Sea for most of the Permian
arid mountains
ZECHSTEIN SEA
N.W. CUVETTE
N.E CUVETTE
Maximum extent of the sea in upper Magnesian Limestone times
MIDLAND CUVETTE
Prevailing wind
Limit of denuded Armorican Mountains
S.W. CUVETTE
Continent of HERCYNIA
VOLCANOES LOCALLY ACTIVE
DESERT LOWLANDS AREA OF DEPOSITION
DESERT HIGHLANDS AREA OF EROSION

It was in this sea, in the Magnesian Limestone Series that some of our important economic rocks were deposited. These are called "evaporites", and are essentially chemicals precipitated on the sea floor as the sea evaporated, and dried out under the hot desert conditions. As the volume of the sea decreased, so the concentration of the salts increased, thus causing chemical precipitation to start. The valuable chemicals in these salts are sodium, magnesium and potassium, which form a substantial portion of the natural materials used in the major petro-chemical industries of the north-east coast, where they are mined. There are three major cycles of evaporite precipitation found in the Magnesian Limestone Series of Yorkshire, which suggests that the sea dried up, and was replenished with more sea water, three times. Such replenishment could be accomplished by the same mechanism which caused the Zechstein sea to form in the first place, such as a periodically subsiding bar leading to a major ocean basin. The salts were primarily percipitated to the east of the limits shown on the map.

Elsewhere, in the low-lying desert cuvettes, the rocks representing the Permian are terrestial dune sandstones and breccias. These are locally important as building stones and as aquifers.

Fossil remains are stunted and rare, being found in few localities. The sea water was too saline for corals to grow, but brachiopods, gastropods, and crinoids are found in the Magnesian Limestone Series.

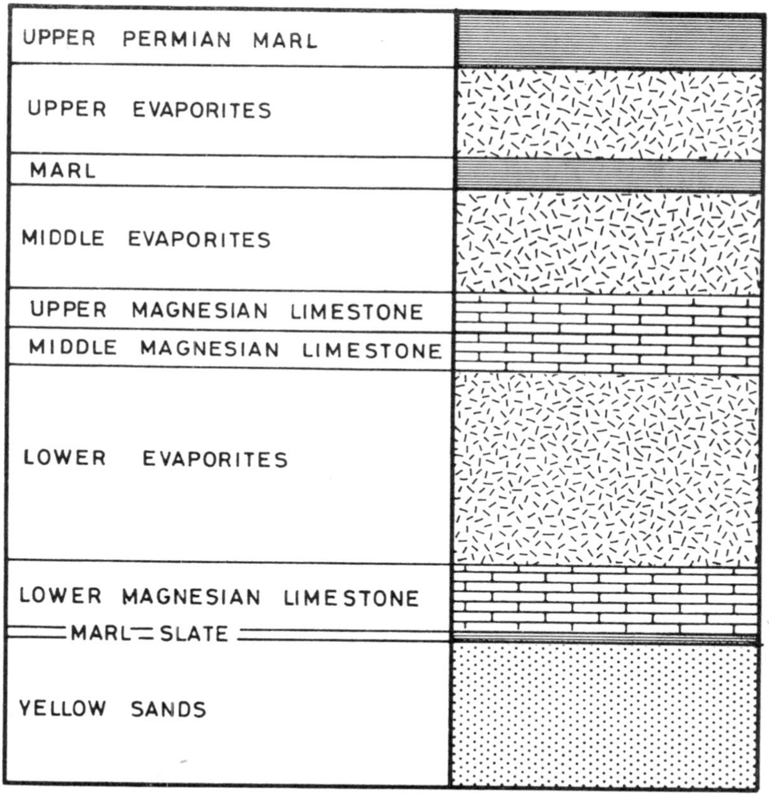

UPPER PERMIAN MARL

UPPER EVAPORITES

MARL

MIDDLE EVAPORITES

UPPER MAGNESIAN LIMESTONE

MIDDLE MAGNESIAN LIMESTONE

LOWER EVAPORITES

LOWER MAGNESIAN LIMESTONE

MARL SLATE

YELLOW SANDS

General sequence of the Permian beds in the north-eastern cuvette. All beds are not present everywhere.

TRIASSIC

Hot desert conditions continued right through the Triassic period, and the Zechstein sea receded from the low-lying desert regions which it had occupied during the Permian. These regions became red desert, and later were the sites of local salt and gypsum-depositing lakes. These deposits are now mined—the salt mainly by pumping steam into it underground and thus dissolving it, and the gypsum by open-cast mining techniques.

The lowest division of the Bunter represents a period of desert conditions in which desert sandstones were formed from the accumulation of sand dunes under extremely arid conditions. Following this, enormous deltas and screes made of sand and pebbles built out into the desert from the Mercian and Welsh Highlands thus forming an intermediate wedge of beds known as the Bunter Pebble Beds. Torrential mountain storms in the highlands caused tremendous rivers of water to flood for short periods into the deserts. These transient rivers eroded, transported, and deposited the sand and pebbles without much sorting—thus large pebbles are found mixed with small sand particles. The pebbles are, however, well rounded. The Bunter sandstones are extremely important as aquifers. They contain and transmit water easily, and are an important source of water in many parts of the country.

TRIASSIC

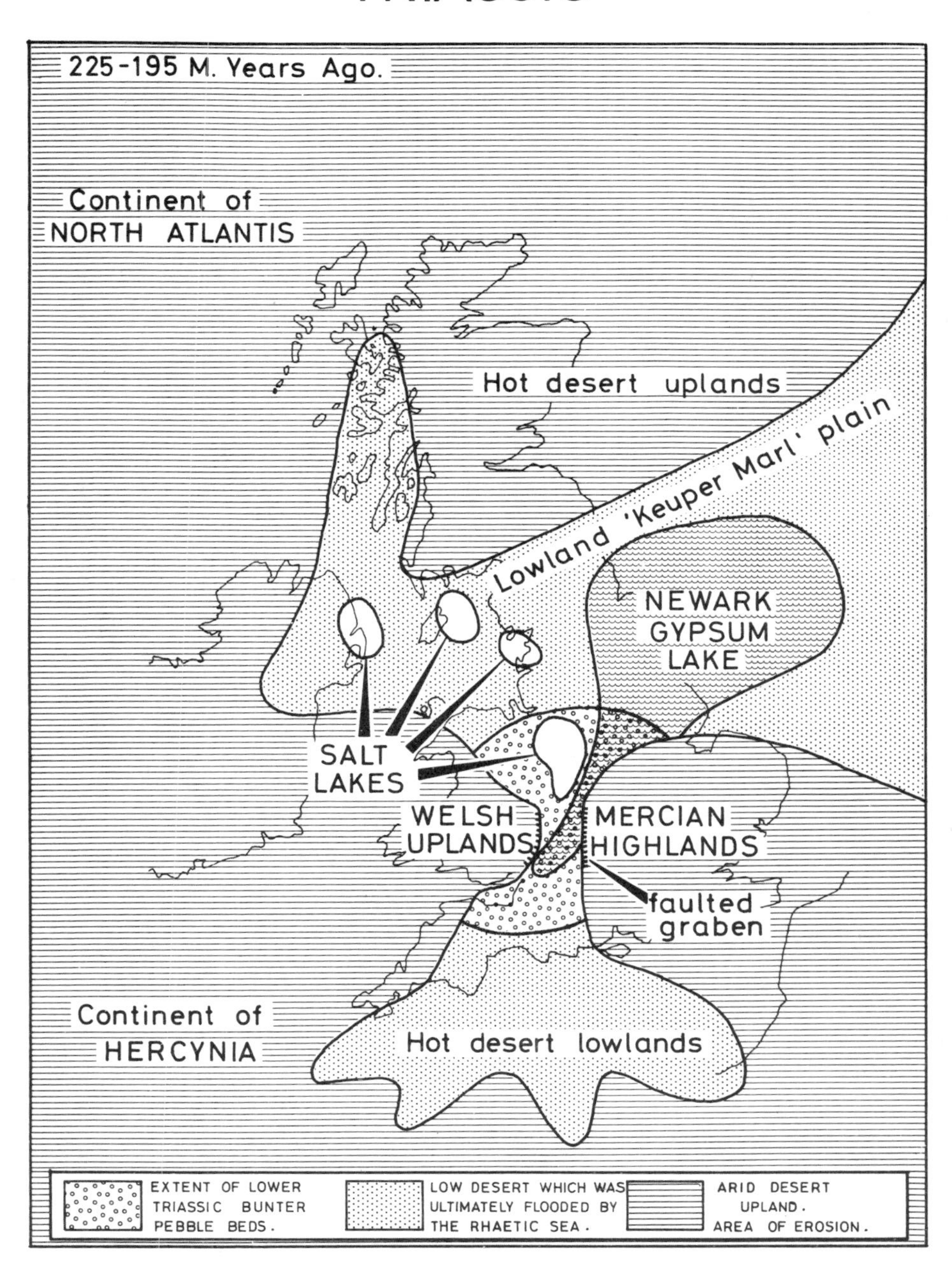
225-195 M. Years Ago.
Continent of
NORTH ATLANTIS
Hot desert uplands
Lowland 'Keuper Marl' plain
NEWARK
GYPSUM
LAKE
SALT
LAKES
WELSH
UPLANDS
MERCIAN
HIGHLANDS
faulted
graben
Continent of
HERCYNIA
Hot desert lowlands
EXTENT OF LOWER TRIASSIC BUNTER PEBBLE BEDS.
LOW DESERT WHICH WAS ULTIMATELY FLOODED BY THE RHAETIC SEA.
ARID DESERT UPLAND. AREA OF EROSION.

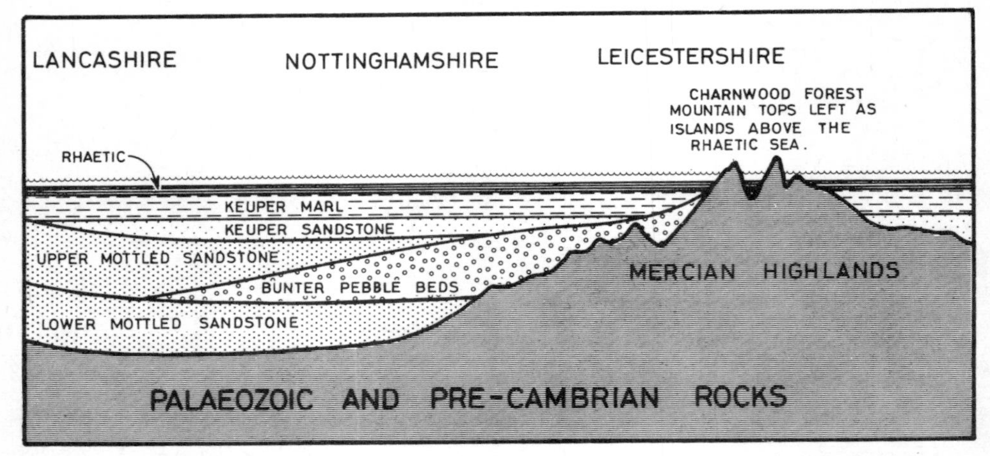

Generalised diagram illustrating the local derivation of the Pebble-Beds from the Mercian Highlands.

The Upper Mottled Sandstone contains sand particles which indicate a desert origin, yet the bedding shows characteristics indicative of shallow-water deposition. This suggests that transient desert lakes had begun to form in the lowland areas.

The Keuper Sandstone was also water-worked, (deposited in shallow moving water), and filled in minor depressions and basins finally to form an extensive plain, on which was deposited the Keuper Marl series.

The Keuper Marl is more extensive than any of the lower sub-divisions, and contains within it the economic salt and gypsum beds mentioned earlier and shown on the map.

The general red colouring of the Permo-Trias rocks is attributed to the oxidising effect of the arid climate on the iron contained in minerals, but at the top of Keuper Marl, the beds are green in colour—a feature taken to represent the effect of plentiful moisture on iron in rock minerals. At that point in time, therefore, the climate suddenly became moist, and shortly afterwards the sea transgressed across practically the entire country. The Rhaetic beds deposited on the floor of the sea are taken as the top beds of the Triassic period. They contain lamellibranch, fish and reptile remains, and are particularly noted for the earliest mammal remains to be found in the British Isles. These are from Rhaetic coastal deposits in the region of the Mendips.

The lack of any igneous activity in the Triassic should also be noted.

TRIASSIC	KEUPER	Rhaetic	MARINE BEDS 15–30 m. THICK SHALES AND MARLS WITH LAMELLIBRANCHS ETC.
		TEA GREEN MARL Keuper Marl	RED CLAY AND SILT BEDS 1000 m. THICK, WITH SALT AND GYPSUM DEPOSITS IN THE LAKES SHOWN ON THE MAP.
		Keuper Sandstone	100 m. OF CROSS-BEDDED WATER-LAID SANDS. FISH, PLANT, AND REPTILE REMAINS.
	BUNTER	Upper Mottled Sst.	BRIGHT RED WATER-WORKED SAND USED INDUSTRIALLY AS A MOULDING SAND. CONTAINS NO PEBBLES.
		Bunter Pebble Beds	YELLOW-COLOURED SANDSTONE WITH QUARTZ PEBBLES RANGING FROM 6–220 mm. DIAMETER.
		Lower Mottled Sst.	RED-COLOURED WIND-DEPOSITED DUNE SANDSTONE. ALSO CALLED THE PENRITH SANDSTONE IN EDEN VALLEY.

JURASSIC

The Jurassic conditions were a continuation of those initiated when the Rhaetic Sea spread rapidly across the British Isles. The climate was sub-tropical, and the warm shelf seas in the area of the British Isles linked two major oceans. The one to the north is called the Boreal Ocean, and the one which was to the south is called Tethys.

There was abundant life in the sea and on land, and the reptiles were a particularly dominant group. Many types of reptile lived on land, while some had become adapted to a marine environment, and others had developed the ability to fly. Because of their large size and slow rate of evolution the Jurassic reptiles are of little use to the geologist for detailed correlation of rock strata. Ammonites, on the other hand, were small in size, numerous, and were evolving rapidly during this period. Consequently, the Jurassic is stratigraphically sub-divided mainly on the basis of ammonite zones. Lamellibranchs and corals were also common.

JURASSIC

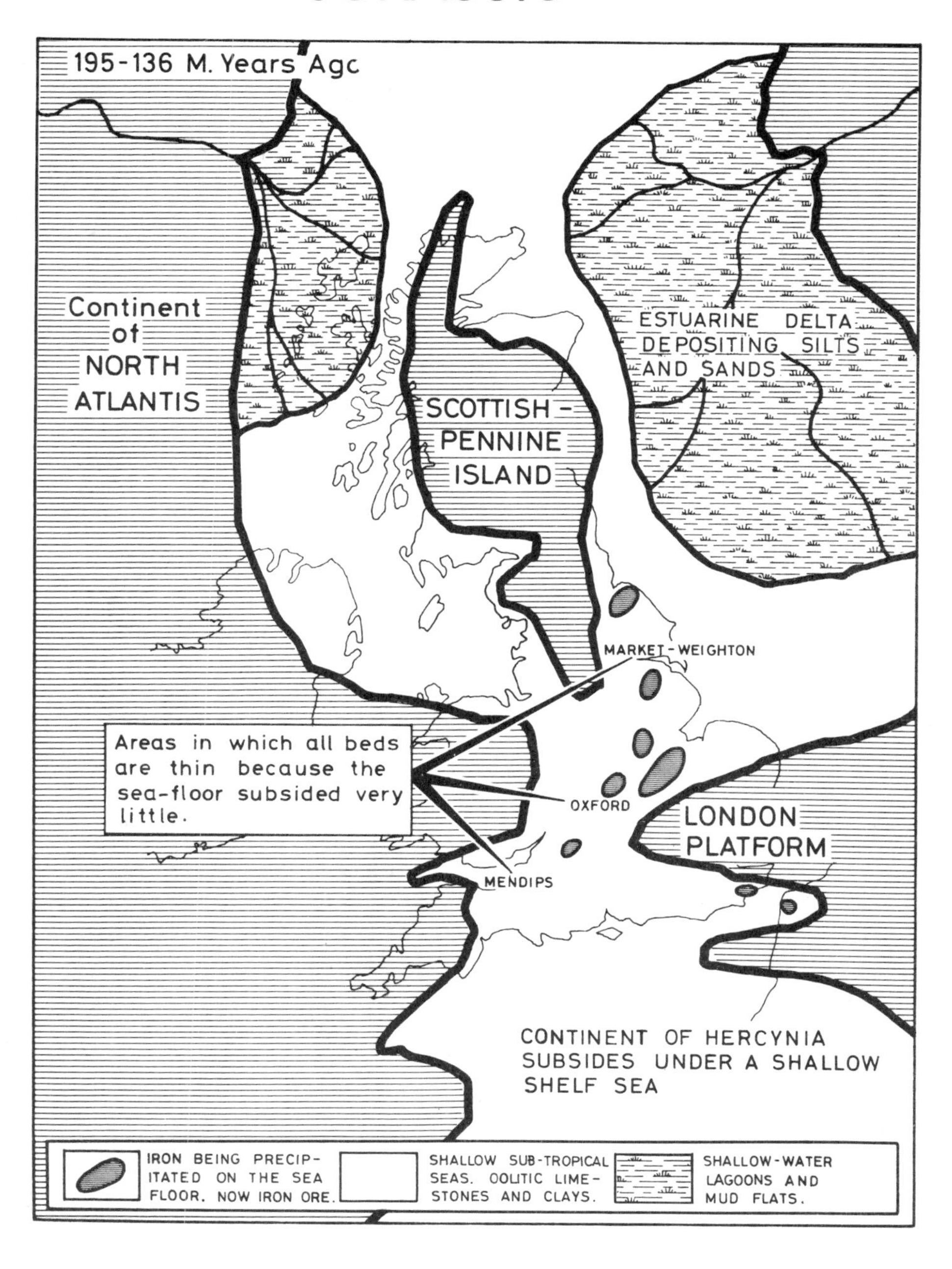
195-136 M. Years Agc
Continent of NORTH ATLANTIS
SCOTTISH - PENNINE ISLAND
ESTUARINE DELTA DEPOSITING SILTS AND SANDS
MARKET-WEIGHTON
Areas in which all beds are thin because the sea-floor subsided very little.
OXFORD
MENDIPS
LONDON PLATFORM
CONTINENT OF HERCYNIA SUBSIDES UNDER A SHALLOW SHELF SEA
IRON BEING PRECIPITATED ON THE SEA FLOOR. NOW IRON ORE.
SHALLOW SUB-TROPICAL SEAS. OOLITIC LIMESTONES AND CLAYS.
SHALLOW-WATER LAGOONS AND MUD FLATS.

Ammonite

Lamellibranch

The map shows that across the country from Dorset to Yorkshire, a stretch of shallow warm sea extended on the floor of which sediments were being deposited. This was not a single sedimentary basin however, but was divided into four minor basins by development of three areas of little subsidence. These areas are sometimes referred to as "axes", and are marked on the map as the "Mendips", "Oxford" and "Market Weighton" axes. Because of the lack of sinking in the axis areas, they are also sometimes referred to as "positive axes". The four basins were as follows: one in the area of Dorset; one in the area of the Weald; one in the East Midlands; and one north of the Market Weighton Axis, in north-east Yorkshire.

This multiple-basin system of deposition meant that beds thickened towards the centre of basins and thinned towards the positive axes. Similarly, the individual rock types and sequences differed from basin to basin and individual names are often necessary for locally developed beds. It is interesting to note that the Market Weighton axis was so positive that only the Lias beds were deposited across it, and even then, only thinly.

Economically, the Jurassic is noted for the ironstone formations which were formed in the warm seas. Conditions were such that iron—in the form of the minerals siderite, chamosite, and limonite—was precipitated on the sea bed. Where the concentrations are economically sufficient, the iron is now worked. These Jurassic ore fields provide most of this country's indigenous supply of iron, and the sites of economic deposits are marked on the map. Across much of the Midlands, the Jurassic rocks outcrop with little or no dip, and where they are near the surface, the ironstones are worked by open-cast mining techniques. The soil overburden is stripped off and retained in dumps. Then the ironstone is quarried. After removal of the ore, the overburden is replaced, and the land returned to farming.

Another economically important rock from the Jurassic is oolitic limestone which makes an excellent building stone. It is different from ordinary limestone in that, during its formation, the tiny shell fragments on the sea floor were washed to-and-fro by oscillating currents and wave action, and consequently had precipitated on them a spherical coating of calcium carbonate. Thus, under magnification, the rock can be seen to consist of an aggregate of tiny spheres cemented together by a calcium carbonate matrix. These spheres are called ooliths, but if they are larger, are sometimes called pisoliths, when the rock becomes a pisolitic limestone.

There is no record of igneous activity during the Jurassic period.

UPPER JURASSIC	PURBECK	MAINLY BRACKISH-FRESH LAKE LIMESTONES.
	PORTLAND	MARINE LIMESTONE. CORALS ABSENT. GIANT AMMONITES ARE PRESENT.
	KIMERIDGE CLAY	MARINE MUDS. UP TO 500m. THICK. CONTAINS REMAINS OF AMMONITES, LAMELLIBRANCHS, BELEMNITES, LARGE MARINE REPTILES, AND DINOSAURS.
	CORALLIAN	70m. OF SANDSTONES WITH CLAY BANDS. AMMONITES, LAMELLIBRANCHS, AND CORALS.
	OXFORD CLAY	MARINE MUDS AND CLAYS. FOSSILS ARE SAME AS KIMERIDGE.
	KELLAWAYS	7-20m. OF CLAYS AND SANDS. ONLY ONE AMMONITE TYPE PRESENT.
MIDDLE JURASSIC	GREAT OOLITE	OOLITIC LIMESTONE. MAIN FOSSILS ARE BRACHIOPODS AND LAMELLIBRANCHS. AMMONITES ARE RARE.
	INFERIOR OOLITE	
LOWER JURASSIC (LIAS)	UPPER LIAS	CLAYS AND SANDS WITH LIMESTONES. AMMONITES, LAMMELLIBRANCHS, FISH, BRACHIOPODS, BELEMNITES, AND REPTILES.
	MIDDLE LIAS	
	LOWER LIAS	

CRETACEOUS

The geography of the Cretaceous changed drastically during the course of the period, and thus two maps are shown in order to describe it realistically.

The main map shows lower Cretaceous times. The land surface had risen above sea level over much of the country, and deposition was taking place only in three of the major subsiding basins which had been established during the Jurassic. These were the Wealden, East Midlands, and North Yorkshire Basins.

In the Wealden area a subsiding, deltaic (non-marine) environment existed on the northern edge of a shelf connected to TethysOcean in the Mediterranean area. In the midlands, shallow marine conditions existed in which sands, ironstones, and clays were deposited. This basin was subsiding slowly, whereas the North Yorkshire Basin subsided rapidly and in it a series called the Speeton Clays was laid down. The North Western Sea was not connected to the Tethys Ocean, but was joined to the Boreal Ocean to the north.

By middle Cretaceous times the two axes had become less positive, and from the south the Gault Clay and Upper Greensand crossed the Oxford Axis. At the same time–for the first time since lower Jurassic–sediments were deposited over the Market Weighton Axis. These were the Carstone, and the Red Chalk. The two seas thus met from the north and south.

CRETACEOUS

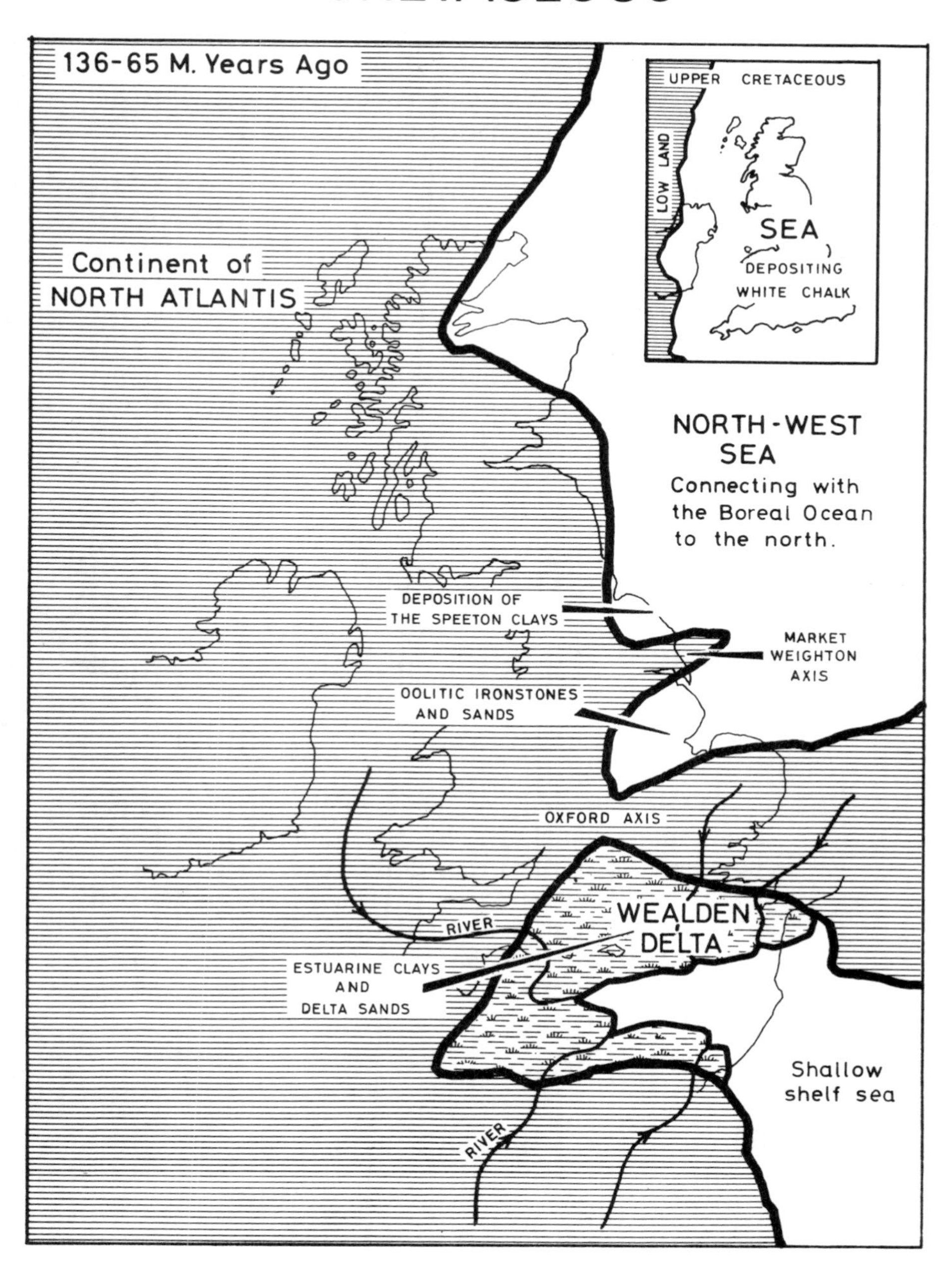
136-65 M. Years Ago
UPPER CRETACEOUS
LOW LAND
SEA
DEPOSITING
WHITE CHALK
Continent of
NORTH ATLANTIS
NORTH-WEST
SEA
Connecting with
the Boreal Ocean
to the north.
DEPOSITION OF
THE SPEETON CLAYS
MARKET
WEIGHTON
AXIS
OOLITIC IRONSTONES
AND SANDS
OXFORD AXIS
RIVER
WEALDEN
DELTA
ESTUARINE CLAYS
AND
DELTA SANDS
RIVER
Shallow
shelf sea

A general submergence followed, and the inset map shows that—with the exception of western Ireland—almost the entire country was covered by the Upper Cretaceous sea. This sea deposited the chalk, and was the last major marine transgression to occur over the British Isles area up to the present day. Chalk is an important economic rock, being widely mined for cement manufacture. It consists of the skeletal plates of microscopic sea organisms, and chemically is calcium carbonate—like limestone, but more pure in that clay, silt, and sand impurities are absent. The lack of sediments in the chalk implies that there were no major rivers bringing sediments into the upper Cretaceous sea, indicating that the adjacent land was probably low-lying with a dry climate.

Belemnite

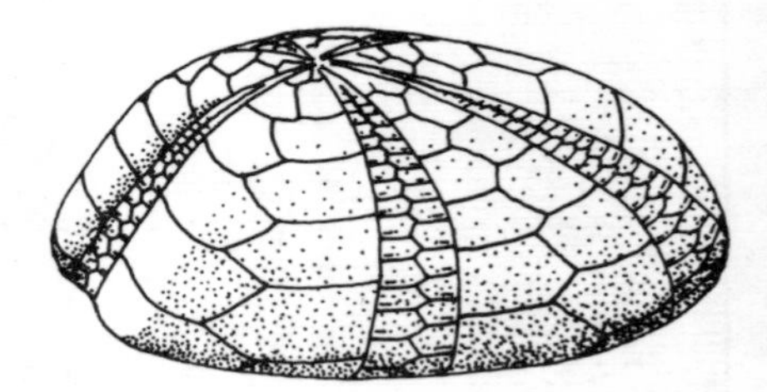

Echinoid

The fossils most abundant in the Cretaceous are ammonites, echinoids, and belemnites, but it should be noted that during the chalk-sea time, the ammonites started to die out, and were entirely extinct by the end of Cretaceous times. Similarly, the dinosaurs became extinct by the end of the Cretaceous.

As in the Jurassic, there is no record of volcanic or igneous activity.

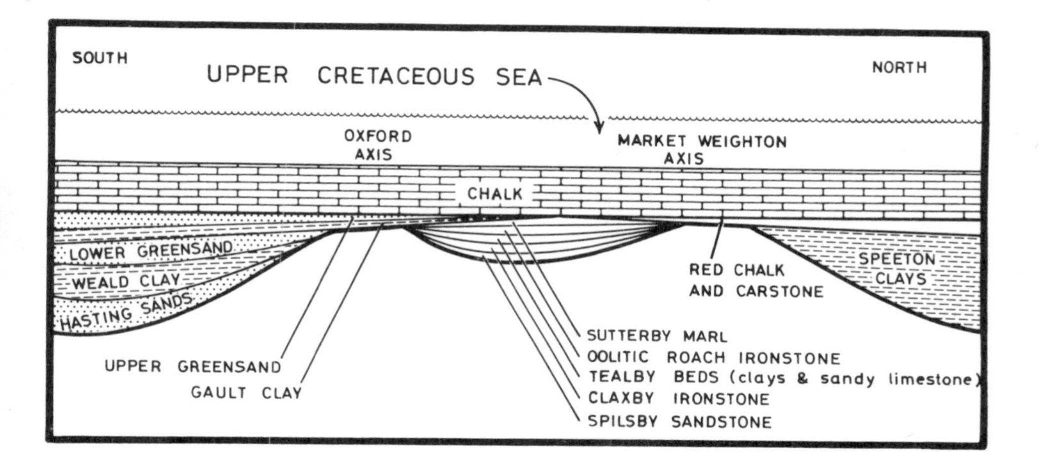
SOUTH
UPPER CRETACEOUS SEA
NORTH
OXFORD AXIS
MARKET WEIGHTON AXIS
CHALK
LOWER GREENSAND
WEALD CLAY
HASTING SANDS
SPEETON CLAYS
RED CHALK AND CARSTONE
UPPER GREENSAND
GAULT CLAY
SUTTERBY MARL
OOLITIC ROACH IRONSTONE
TEALBY BEDS (clays & sandy limestone)
CLAXBY IRONSTONE
SPILSBY SANDSTONE

TERTIARY

In early Tertiary times sands and clays were laid down in a basin extending from East Anglia to Dorset, but by mid-Eocene times, the area of deposition had contracted, and subsequent beds of late Eocene and Oligocene age were only laid down in the Hampshire area.

During the Oligocene, major mountain building was in progress in other parts of the world as the Alpine Orogeny established itself. It was not, however, until early Miocene that any tectonic effects were felt in the British Isles. The beds laid down during the Eocene and Oligocene were now lifted out from the sea and folded to form the London syncline, the Wealden anticline, the Dorset anticline, and the connected Isle-of-Wight monocline. The major structural blocks of the British Isles were also uplifted at this point to their present elevations. This includes the Scottish Highlands, the Southern Uplands, the Pennine Chain, the Lake District dome, the Welsh Mountains, etc. The tectonic pressure from the south also caused the formation of a series of dextral tear faults in the region of Cornwall and Devon, and caused substantial movement along most of the pre-existing fault systems in this country, such as the Church Stretton Fault and the Pennine Fault System.

Another important feature of the Tertiary was the development of the North-West Igneous Province in the region of the present North Atlantic Ocean.

TERTIARY

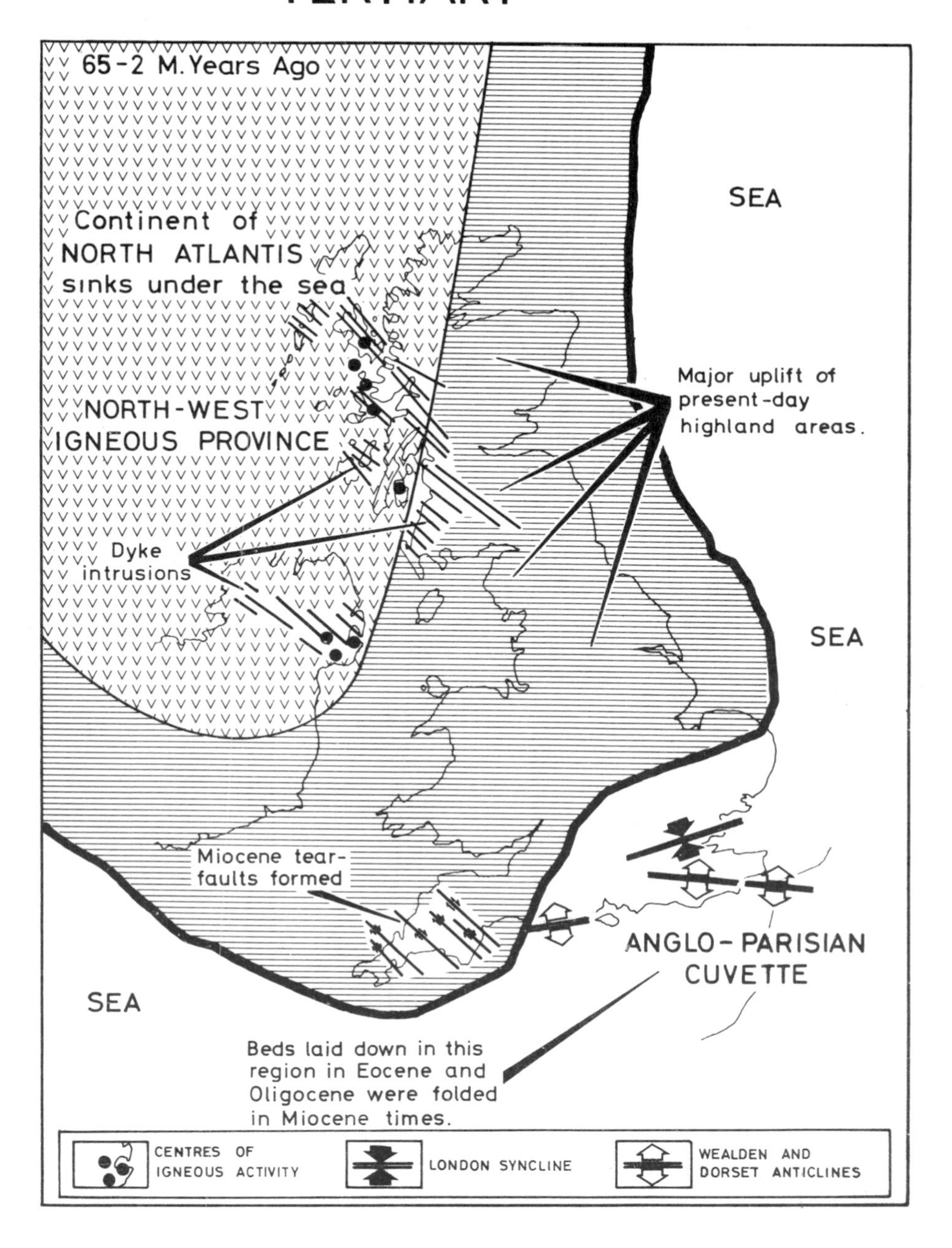

65-2 M. Years Ago
SEA
Continent of NORTH ATLANTIS sinks under the sea
NORTH-WEST IGNEOUS PROVINCE
Major uplift of present-day highland areas.
Dyke intrusions
SEA
Miocene tear-faults formed
ANGLO-PARISIAN CUVETTE
SEA
Beds laid down in this region in Eocene and Oligocene were folded in Miocene times.
CENTRES OF IGNEOUS ACTIVITY
LONDON SYNCLINE
WEALDEN AND DORSET ANTICLINES

During early Tertiary times, marked outbreaks of igneous and volcanic activity took place in the region shown on the map. Combined with the widespread intrusion of large igneous masses of granite and gabbro, were tremendous outpourings of basaltic lava which formed a lava field over millions of square kilometres. Sills and dyke swarms were injected into the country rocks, with a predominant dyke-swarm direction of north-west to south-east. The culmination of this intense igneous activity came in the late Tertiary, when nearly the whole of the North-West Igneous Province foundered and sank below the sea, thus forming the northern part of the North Atlantic Ocean. Remnants of the original Province are found in the Antrim plateau basalts of northern Ireland, and the igneous intrusive centres of the Western Isles of Scotland.

Lamellibranchs, gastropods, corals and echinoids are very important fossils from the Tertiary system. It should be noted that, although not used as zoning fossils in this country, mamals became numerous during the Tertiary, and also evolved rapidly.

		HAMPSHIRE AREA	LONDON AREA
TERTIARY	PLIOCENE	ABSENT	ABSENT
	MIOCENE	ABSENT 20 MILLION YEARS OF FOLDING AND EROSION. NO DEPOSITION.	ABSENT
	OLIGOCENE	HAMSTEAD BEDS	ABSENT
		BEMBRIDGE BEDS	
		OSBORNE BEDS	
		HEADON BEDS (upper & middle)	
	EOCENE	(lower)	
		BARTON BEDS	
		BRACKLESHAM BEDS	
		BAGSHOT BEDS	BAGSHOT BEDS
		LONDON CLAY	LONDON CLAY
		READING BEDS	READING BEDS
		ABSENT	THANET SANDS
	PALAEOCENE	ABSENT	ABSENT

ALPINE MOUNTAIN BUILDING PERIOD

EFFECTS OF ALPINE OROGENY REACH BRITAIN

ALPINE MOUNTAIN BUILDING ACTIVE IN MEDITERRANEAN AREA, BUT NOT AFFECTING BRITAIN.

QUATERNARY

The Quaternary is sub-divided into the Pleistocene (most of the Quaternary's two million years) and the Holocene (the last ten thousand years up to the present day).

The main geological features of the Pleistocene are those resulting from tremendous climatic changes, and include evidence of the only glacial activity that this country has known in the last six hundred million years. Although the period lasted for only two million years, during this time the world's climate became extremely unstable and fluctuated rapidly. This resulted in the periodic cooling of the British Isles with consequent ice-sheet formation. Glaciers eroded the highlands forming scenery with characteristics such as "U"-shaped valleys, corries and hanging valleys, while the lowland areas were smeared with the erosion debris and boulder clay derived from the uplands.

In the British Isles, three major periods of glaciation are recorded. (First: Lowestoftian; second: Gippingian; third: Weichselian.) These reflect three major deteriorations in climate, of which the first two were much more severe than the last. In other localities such as the Alpine areas, earlier glaciations have been recorded, but evidence for these is absent in the British Isles. The only beds of this age in the British Isles are cold-water sedimentary deposits which accumulated in a sinking basin in the coastal

QUATERNARY

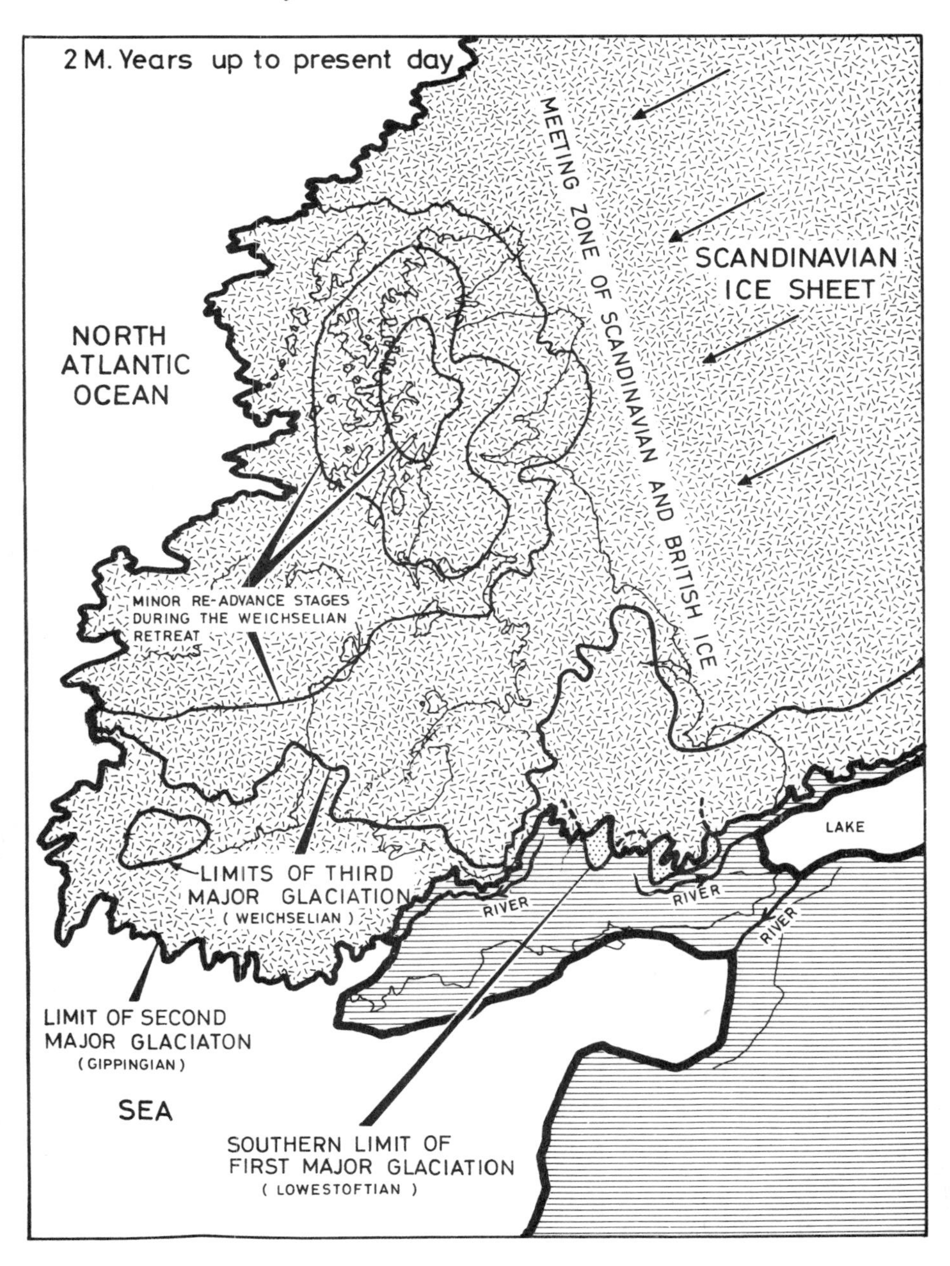
2 M. Years up to present day
MEETING ZONE OF SCANDINAVIAN AND BRITISH ICE
SCANDINAVIAN ICE SHEET
NORTH ATLANTIC OCEAN
MINOR RE-ADVANCE STAGES DURING THE WEICHSELIAN RETREAT
LIMITS OF THIRD MAJOR GLACIATION (WEICHSELIAN)
LAKE
RIVER
RIVER
RIVER
LIMIT OF SECOND MAJOR GLACIATON (GIPPINGIAN)
SEA
SOUTHERN LIMIT OF FIRST MAJOR GLACIATION (LOWESTOFTIAN)

region of East Anglia. These constitute the Beestonian earlier beds.

The first of the three glaciations for which there is evidence is called the Lowestoftian. Exposed Lowestoftian boulder clays (or tills) are limited in area and are shown on the map; they are mainly covered over by the Gippingian (second glaciation) deposits. The Weichselian glaciation was not as extensive as the first two, and even much of the Midlands was not covered in ice during this time. The Quaternary beds, therefore, constitute a series of three groups of boulder clays separated by boreal and temperate sediments with fossiliferous beds containing especially mammals and even early Man. South of each of the main boulder clay sheets are found moraine and outwash gravel deposits.

The marked instability of the climate is demonstrated by the evidence for warm periods in between the major ice advances. Such evidence consists of fossilised plants and animals which are only found today in warmer climates. In some of these periods—called interglacials—the temperature became warmer than the present day, but there were also many minor ameliorations of climate called interstadials, during which the climate never became as warm as the present day, but could support tundra, plants and coniferous forests.

At the beginning of a period of glaciation, the temperature would fall, and individual ice sheets would build out from local highland centres such as the Highlands of Scotland, the Southern Uplands, the Lake District, etc. These ice sheets would meet and spread out over the lowlands, thus creating a virtually continuous cover of ice. Such a massive ice sheet also spread out towards the British Isles from Scandinavia, and encountered our ice sheets in the area shown on the map. The ice sheet in Scotland became so thick that the crust in this region was depressed isostatically, and at present is rising since the weight of the ice has been removed. Consequently, the Scottish coastlines are being lifted out of the sea, and show features such as raised beaches and land-locked fiords.

About eight thousand years ago, the sea finally spread across the Isthmus of Dover to form the Straits of Dover, and thus truly form the British Isles as we know them today.

BRITISH SUB-DIVISIONS					EQUIVALENT SUB-DIVISIONS	
					ALPINE	N. EUROPEAN
QUATERNARY	HOLOCENE			RECENT GRAVELS AND SOIL DEPOSITS		
	PLEISTOCENE	UPPER	FLANDRIAN			
			WEICHSELIAN	◄ GLACIATION ►	WÜRM	WEICHSELIAN
			IPSWICHIAN			
			GIPPINGIAN	◄ GLACIATION ►	RISS	SAALE
		MIDDLE	HOXNIAN			
			LOWESTOFTIAN	◄ GLACIATION ►	MINDEL	ELSTER
			CROMERIAN			
			BEESTONIAN	DEPOSITS OF THIS AGE ARE FOUND ONLY IN E. ANGLIA	GÜNZ ◄	GLACIATION IN ALPS WITH NO BRITISH EQUIVALENT
			PASTONIAN			
		LOWER	BAVENTIAN			
			ANTIAN			
			THURNIAN			
			LUDHAMIAN			

IT SHOULD BE NOTED THAT VERY RECENT NOMENCLATURE IS DIFFERENT FOR THE MAIN U.K. GLACIATIONS WHICH ARE CALLED AS FOLLOWS UNDER THE NEW SYSTEM:
WEICHSELIAN = DEVENSIAN; GIPPINGIAN = WOLSTONIAN; LOWESTOFTIAN = ANGLIAN.